Eclipse *i-clipse, n,* the total or partial disappearance of a heavenly body by the interposition of another between it and the spectator, or by passing into its shadow.

Chambers Dictionary

Copyright

©

JP Gough

© 2021 JP Gough.

get in touch

contactgough@gmail.com

The Eclipse
of the modern mind

A TRILOGY

First Contact

CONTENTS

Eclipse

{diagram}

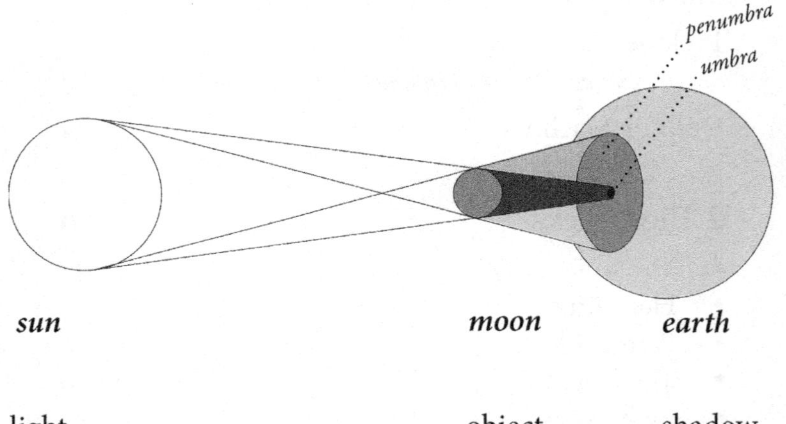

Phases of the Eclipse

First Contact

{this book}

Shadow approaching: the observer enters the penumbra, the start of the partial eclipse as object starts to impinge on light source: the moon's dark shadow grows from the West.

Second Contact

Shadow moves toward completion: the umbra is now projected at the observer's location and totality begins as object moves completely within the light source: the Western Moon covers sun.

Totality

Shadow turns to temporary night: the moment (in minutes) darkness is complete and object completely covers light source within the umbra of observer.

Third Contact

Shadow moves out and Totality ends, the observer exits the umbra and sunlight returns from the West.

Prologue

The Problem of the Shadow

I have a projector in my study. It is fixed to the ceiling behind my desk. That I would end up sitting for hours in front of it observing my shadow on a screen, noting down the philosophical implications, was not really what I had in mind when I bought it. It's supposed to be used for flight simulation – one of my pet obsessions, along with this book, which has been years in the making. And to think after all these years, after all the reading, all the research, all the thinking, it would all end up with me staring at my shadow.

And considering the implications of this my latest craze, I don't think it comes as any surprise if I have come to the conclusion that fact really is stranger than fiction. For stranger am I, and stranger I have become in pursuit of my shadow so as to solve THE PROBLEM. As you can imagine it is rather a deep one, devilishly complicated, obscure. Can't say I care for it much, because from the 'outside', as my wife will tell you, it's nuts. On his chair 'studying it' again. (Not that she takes much notice, she's used to IT). But from the 'inside' I assure you IT IS different ... a curious encounter with the Shadow .. The Eclipse of the Modern Mind .. a strange book about aliens spaced out in the darkness of the other side. On a strange mindless odyysey, they go in search of a truth they cannot figure out, can't quite find. Who knows, perhaps they will one day they will, but one thing's for sure and I might as well say it now. I have. Figured it all out, I mean. Can't say it's very complicated. Bit obvious really. Because clever and sophisticated as they are, however hard they try and find the truth 'out there' in the parenthesis of the shadow (ie. the meaningless absurd cosmos we all live in) they are not going to do so. Bit silly really. And I must say I am satisfied. I have done it! And it has all been worth it, because my search is over and I have found THE SOLUTION, the answer to the question: why is our world so so f**ked up? It's been my life mission, and if you are anything like me, confused and perplexed by it all – well, welcome to the shadow.

And by the shadow I mean this. Take any object, any body or thing you care to think of. Now the eclipse. Hold it up to the light so as to cast a shadow. Let this light be the light of science. As it encounters the object so all its qualities are occluded, annihilated, to leave a shadow. What remains, what is real, is given to quantity, now the mass, substance of the shadow, now the very 'cause' of the object-world so upheld. Truth now corresponds to the shadow. Sound crazy? Welcome to the metaphysical foundations of modern science and some astonishing magic: the exchange of light for darkness, a whole world pulled inside out like a rabbit from a hat. But how deep does the rabbit hole go? What is this bloody shadow, not distant, far and remote, but haunting us here in skull and bone? For now we see in a mirror darkly: face to face with the shadow. But how did it THAT get there? And how did we end up playing this ghostly far-fetched game of hide and seek? How were we cast down in the doom and gloom, end up so lost confused in the darkness of the modern ego? Desperate to find a way out, we know not how, have not the remotest idea of how we came to be here.

And when all this is clinically and most certainly certified as corresponding to 'reality' and 'normality' and all the rest by that highest of the highest understanding known as Modern Science – then frankly we have a problem: a problem of premises, the premises of this the earth we live on and call our home.

Question being – are we all completely MAD?

For you may cast a shadow, just like anybody or anything naturally casts a shadow. We know perfectly well what it is. You are obviously not it, most certainly not it, not possessed by it, nor anything remotely like it by any stretch of the imagination, nor is the world we live in any suchlike 'shadow state'. (Or is it?) Because between the two lies a vast difference. And in case this needs further explanation, let us digress into those metaphysical foundations. Here's the difference. One is natural, the other esoteric and abstract – so abstract as to become the spiritless fancy of the minuscule mind of what – an insect perhaps? Because in its unhinged state, this mighty little tick of deranged logic has somehow managed to make up its mind in such a way as to turn itself inside out and the other way round so as to cast itself on the other side of itself – like a shadow. For who would be so taken in so as to believe it's true? That you are it, and it is you? And not just you, but every body and thing, the whole universe, not mention a little planet called earth? Welcome to the eclipse of the modern mind. Nor is it difficult, complicated or obscure. Hmm, now how does one get out of it ...

Durr.

Mind you, when I was reading Plato's **Allegory of the Cave** again for the umpteenth time and the idea of 'the shadow' first popped up in my mind, I immediately dismissed it as being too facile. Doh. Cannot be that simple. Then, later, I came back to it and was astonished by its blinding logic, sheer precision and eerie correspondence to all the ills of modern civilization. It was too good to be true – or rather too bad, considering the implications. How the hell could the Western Mind have made such a confounded error, monumental mistake? And in the process of rewriting my original manuscript, the further I got the more dumbfounded I became. It all seemed to fit, as slowly, step by step, chapter by chapter, I found myself under increasing duress and forced to submit to an unspeakable truth that I dared not tell, let alone publish, for fear of ridicule.

But I must and I will. Because I can't fault it. But if you can, do tell. It's taken me a while, but every minute of every hour has been worth it, because now I have the confidence, can verily say and happily proclaim my innocence. Because once I was guilty of the construct, felt most unworthy of the shadow, but now it is not worthy of me to lower myself to the shallow superficiality of such a foolish notion, catastrophe of reason – so as to proclaim my release from the prison, the redemption of my soul, and not only this, but the saving of a planet caught in an eclipse. Even an idiot can see it. Fool no more. Come to your senses. Witness the predication of the eclipse, open your mind, allow the epiphany to come. Let it happen, let it dawn, so sunlight may return. That's another chapter, another book, one that perhaps need not be written – I mean, do I really have to tell you how easy it is to be merrily released, no longer stuck in the shadow?

But for now, let's get in the mood and find somewhere quiet where we can focus on the task in hand, somewhere where we may sit against a wall so as to observe the shadow, and when the wall becomes the world as mirrored in the construct, so we witness a magical most metaphysical event: the eclipse of planet earth. And here we are changed, come aliens come from afar, as if from another galaxy, another star, some remote place where the shadow reigns, the sun never shines and it is always dark. Here, come what may, truth will never see the light of day, and life is just shades of grey.

Because the writing is on the wall, and there it is, the shadow, so as to reveal the Eclipse: a world gone mental without the remotest idea, hope in hell. Be it fact or fiction, it doesn't really matter – doesn't make a blind bit of difference. Because however hard we try to 'figure out' the truth we never will, because it would be just another shadow. And anyway, how can a shadow judge shadow, make any value judgement at all? But when an idiot points it out to you as (considering the premises) idiot he must be to see it and you to believe it's true, then at least as idiots we shall all be united in that army of degenerates who at least stand a chance of redeeming our humanity by not so much as saving our planet but exchanging it for another one: the one I live on, the one you live on, the one that's not dark, downcast and made over to the status quo, but given to a vision that is the birthright of every soul. Some call it Religion or Natural Morality, others Reason, or just plain Common Sense, but whatever it is, one thing's for sure: it is not a shadow, nor are we obliged in any shape or form, to live under it.

For it seems to me, when it comes down to it, there will no truth by which we are remembered in the apocalypse to come (technically called Totality), save that which we surrendered to the powers that be on such a planetary scale so as to predicate the eclipse. In the last analysis, this will be the 'extraordinary fact' of the child of posterity: how we let it happen, let ourselves be so blindfolded as to walk over the abyss. For how DID they come to rule and measure such folly over us?

Unless of course it dawns upon us, the very idea. Only by predication do we derive logic, and no fact is stranger than the logical predication of what we all know to be happening. It is called an eclipse. Of a particular and peculiar kind, strange indeed, something that happens for NO REASON at all: AN ACCIDENT WAITING TO HAPPEN. Hiding under the shadow is not going to make it go away, nor is the banging pots and pans of extinction rebellions, or the fear and loathing of the construct. For our eclipse is of the human mind, one in kind and indissoluble with the soul, so as to give reason to love, a foundation for hope, not to mention free will. Which means we have a choice about which way the eclipse goes. One way is darkness, the other is light. For God is, and we are not (considering the premises) and yet we are co-created and redemption is at hand for those meaningless accidents of the shadow who were doomed from the start. This not so much a reformation or restoration, rebellion, even revolution – but the RESURRECTION – as all things considered, resurrection it must be – of our consciousness, something not so much willed as realized, something different in kind – the breaking of light, coming of dawn.

> Now more than ever, so it seems, we have all the time in the world to sit back and reflect.
> Woken indeed.
> For who would gain the world and lose their soul?
> And who wants to be a slave of the system?
> Live under a shadow?
> Beats me.

Introduction

When the Artemision Bronze, a magnificent bronze statue of Poseidon, was lifted out of the Aegean sea, it had only hollow sockets for eyes, eyes that had once been gems, intricately crafted and beautifully inlaid, eyes now lost to the ocean, never to be seen again.

This book is the story of those precious gems and how they were lost, gems that once belonged to a spiritual vision of the cosmos until removed from their sockets. If you watch an eclipse, it's a good analogy as to how this process came about – as slowly, imperceptibly, the sun is blinded, now that hollow socket, vacant gem.

For us who live in objectivity, us who live the eclipse, us who see reality in the 'object-light' – through 'objectivity' – by which we mean we see things materially as objects so as to interpose, occlude consciousness, it hardly ever occurs to us – actually it is almost impossible for us – to appreciate that the human word is the *intellectus in actu*, that essentially you are what you think, and actually, a lot more than you think, should we adopt a somewhat Medieval attitude. Like to us, the sun was *intellectus in actu* – the Divine Soul, the Intelligence, that literally lit up the mind like a lantern, thus to reveal the world. When the once, now debunked, natural philosophers talked about the 'Word' coming forth from the Divine Intelligence so as to make up the reality of the world around us, they meant it. The *intellectus* was being realized in *actu*. For them, what existed 'out there' is not already there before you see it, nor is it set in stone, because matter is not fundamental, and nor are you a passive observer. Far from it. What exists 'out there', only exists in 'potentia', until apprehended by the mind in the act of perception.

Hence their obsession with form rather than matter, and the whole movement of their minds was contrary, in the opposite direction to our own. The light of consciousness came from within and moved outwards in what was a co-creative process, for we were 'God-like', literally 'God-seeing', and the Soul was the immanent and transcendent principle uniting us all in an organic cosmos. For each part participated in the whole, in the creative Intelligence. Words were not mere descriptive things, nor was thought confined to the brain: rather words were the immanence and transcendence of the intellectus in actu as it brought forth the world.

And not only did they mean it, they lived it, they breathed it, this Divine cosmos illumined by the Soul. And we may note that this notion is not entirely alien to modern sensibilities, for by the time we have removed the strangeness from quantum physics and its buzzwords of consciousness, we might find we are left with the Scholastics, and those Medieval philosophers, so easily dismissed, turn out to be not quite so ignorant as we thought.

But taking out the eye may be one thing, putting it back in is another. For it might not occur to us that we are what think, but even if we did, we would probably miss the point, clumsy and literal minded as we are, so accustomed to the idea, that thinking somehow goes on in the hollow socket of our brain. The optic nerve after all, is about the size of a pinprick. Try threading that soul connection through the eye of materialism. For ours is a delicate operation. And even if we did see the light at the end of tunnel, we would never have the nerve, daresay even grasp it – the very idea. For us, it is so difficult, damned hard to see, nor are we used to the delicacy of feeling, subtlety of thought. For others not so. To them it was natural. They had the moral fibre, were actually formed by the sinew of Spirit, were already connected to Nature, nor did they perform that surgical operation called the Scientific Revolution so as to eliminate original participation.

Forged by atoms and genes and molecules, bronzed by materialism, we moderns, cold and sophisticated as we are, know we stand as much chance of retrieving those eyes as the Artemision, which now stands in the Athens museum. For those eyes are irretrievable, lost forever. And even if we did manage to replace them with our own dazzling gems of sophistication, we would only be adding insult to injury: they would be fake artificial imitations, useless ornaments,.

The statue after all, is just a statue, and it cannot see any more than we, stuck as we are in our objectivity, occlusion of consciousness. For we have no ears to hear, and however much we look we will never see. For our minds are dull, our ears deaf, our eyes blind, just like the statue, which cannot really hear or understand anything. It just stands there in blind materiality, has not the remotest idea.

Unless of course the statue came to life. Then it would simply grow back its eyes, smash the glass and walk out the cubicle. No longer captive to matter, no longer Artemision, no longer bronze: for no longer would matterkind make over unto its likeness whatsoever came in touch with it. Nor would it have to go in search of anything: for what it was looking for was found within. From the dark sightless pool of nascent consciousness comes a surprise, a miracle, a hope beyond all expectation. In the darkness, something happened, something occurred – maybe, just perhaps, the statue stirred, a thought crossed its mind, something dawned: the hands that made it are just like it's own, as are the ears that hear, the eyes that see, much like the mind of which it was made. And from that touch of genius, sublime thought, comes a vision, the seed is sown, something wakens, breaks the mould. Astonished, the eye grows, begins to glow, no longer bronzed, immobile, but made of a different metal: flesh and bone, blood and spirit, quite human, actually divine: born of a potential beyond its wildest dreams, so redeemed by the Word, the lifeblood, the intellectus in actu.

For if you are what you think, one should be very careful what one thinks: for one day you may find yourself not so much in the modern museum of breaking bronze and smashing materialism, but somewhere else: in a new kind of awareness, felt of as a sort of glow, delicacy of feeling, reconnection of the Soul: the resurrection of yourself, as a new heaven and earth are formed – literally – before your eyes: a dawning, waking of consciousness.

But as to our former state, how can we describe this? How were we before? What analogy, what metaphor can we bring forth? Those sightless pools, that vacant stare, the statue, motionless – the abyss, the occlusion of consciousness, those sleepless sockets – now where have we seen those before?

In Plato's Allegory of the Cave and those prisoners staring at shadows on a wall, called as they are to witness the eclipse of their world.

But Plato's Allegory as never told before, as the Cave becomes the Matrix of our modern understanding. And origin of this understanding is the subject of this book as we introduce the first of three phases of the eclipse. Each phase is technically called a 'contact', and our beginning is the First Contact, which appears to be sometime around 400 BC, roughly the same time the statue of Poseidon was knocked off its perch and thrown into the sea. Starting with the Greeks, we going to watch our mortal frame fall into the ocean of materialism. And then, after resting for centuries in darkness, we lift the statue out of the depths so as to reveal it for what it is. For holding it up for all to see, we might just then realise the folly of the exercise, might possibly wake up to the fact we are sleepwalking into the abyss. For those hollow sockets, they are ours: formed by the eclipse.

The Eclipse of the Modern Mind

The
First
Contact

A trilogy based upon Plato's Cave

THE ALLEGORY OF THE CAVE
In the cave, which is a symbol of human life for most, prisoners watch ...
UNIVERSE INSIDE YOU IS THE CAVE!
... the spectacle of shadows put on by puppeteers.
Life in the shadows
Thinking the shadows are real The prisoners give each other prizes to those who analyze them best and remember their sequence.
underground
He turns and is blinded briefly By the light of the fire used to cast the Shadows.
Intuition tells one man that The shadows are illusions and ...
But he sees the puppeteers And their objects and so thinks them real.
He must break FREE!!!
Intuition tells him as he moves past
SCIENCE FOR LIFE
THE CAVE
THE FIRE
THE ROADWAY
DIFFUSED DAYLIGHT
THE ROUGH ASCENT TO SUNLIGHT
COOL!
WOW!
THIS ONE REALLY GETS THEM GOIN'!
THE SUN
truth
PERFECTION
WINNER
REAL LIFE
Then he notices the objects and realizes That he is gazing on The Real World.
With this knowledge he knows he is completely Free and returns to the cave to set the others free.
shadow
prisoner
reborn NOW
Then his eyes glimpse the Sun and he knows Instinctively that this is Ultimate Reality, The Source of all things and that this light Is a symbol for his own consciousness, that this is the source of his reality.
BELIEVE
ILLUMINATE
They may not accept his new found enlightenment.
They may call him mad. They may even put him to Death. In spite of this he has to try.

Welcome to the

First Contact

A world which begins with Plato

And the most famous passage

in greek philosophy

•

The allegory of the cave

•

Theme of this book

Matrix of our World

Imagine prisoners in a cave forced to face a wall. Up and behind there is a parapet, a stage with a fire over which objects are held by bondsmen so as create shadows below. The prisoners, chained from birth, come to believe the shadow world is real because it is all they know, until one wakes up to escape from the cave and return to tell of what he has seen.

IF THE DOORS OF PERCEPTION WERE
CLEANSED EVERY THING WOULD APPEAR
TO MAN AS IT IS: INFINITE. FOR MAN HAS
CLOSED HIMSELF UP, TILL HE SEES ALL
THINGS THRO' NARROW CHINKS OF HIS
CAVERN.

William Blake

●

PLATO'S ALLEGORY OF THE CAVE

Picture this.

Says Socrates to Glaucon, in Plato's famous Allegory of the Cave.

"Men dwelling in a sort of subterranean cavern with a long entrance and open to the light on its entire width. Conceive them as having their legs and necks fettered from childhood, so that they remain in the same spot, able to look forward only, and prevented by the fetters from turning their heads. Picture further, the light from a fire burning higher up and at a distance behind them, and between the fire and the prisoners and above them, a road along which a low wall has been built, as the exhibitors of puppet shows have partitions before the men themselves, above which they show puppets."

"All that I see." Said Glaucon.

"See also, then, men carrying past the wall implements of all kinds that rise above the wall, and human images and shapes of animals as well, wrought in stone and wood and every material, some of the bearers presumably speaking and others silent.

"A strange image you speak of," said Glaucon, "and strange prisoners."

"Like to us," I said. "For, to begin with, tell me — do you think that these men should have seen anything of themselves or of one another except as the shadows cast from the fire on the wall of the cave that fronted them?"

"How could they," said Glaucon, "if they were compelled to hold their heads unmoved through life?"

"And again, would not the same be true of the objects carried past them?"

"Surely."

"If then they were able to talk to each other, do you not think that they would suppose that in the naming of the things that they saw they were naming passing objects?"

"Necessarily."

"And if their prison had an echo from the wall opposite them, when one of the passers-by uttered a sound, do you think they would suppose anything else than the passing shadow to be the speaker?"

"By Zeus, I do not." Said he.

"Then in every way such prisoners would deem reality to be nothing else than the shadows of artificial objects."

"Quite inevitably," he said.

"Consider then, what would be the manner of the release

and healing from these bonds and this folly, if in the course of nature something of this sort should happen to them. When one was freed from his fetters and compelled to stand up suddenly, and turn his head around and walk and lift his eyes to the light, and in doing this felt pain and, because of the dazzle and glitter of the light, was unable to discern the objects whose shadows he formerly saw, what do you suppose would be his answer if someone told him that what he had seen before was all a cheat and an illusion, but now, being nearer to the reality and turned toward more real things, he saw more truly? And if also one should point out to him each of the passing objects and constrain him by questions to say what it is, do you not think that he would be at a loss and that he would regard what he formerly saw as more real than the things pointed out to him?"

"Far more real." Said Glaucon.

"And if he were compelled to look at the light itself, would that not pain his eyes, and would he not turn away and flee to those things which he is able to discern, and regard them as in very deed more clear and exact than the objects pointed out?"

"Of course."

"And so finally, I suppose, he would be able to look upon the sun itself and see its true nature, not by reflections in water, or phantasms of it in an alien setting, but in and by itself in its own place."

"Necessarily."

"At this point he would infer and conclude that this it is, that

provides the seasons and the courses of the year, and presides over all things in the visible region, and is in some sort the cause of all these things that they had seen."

"Obviously," he said, "that would be the next step."

"Well then, if he recalled to mind his first habitation and what passed for wisdom there, and his fellow bondsmen, do you not think that he would count himself happy in the change and pity them?"

"He would indeed."

"And if there had been honours and commendations among them which bestowed on one another, and prizes for the man quickest to make out the shadows as they pass, and best able to remember their customary precedences, sequences, and co-existences, and so most successful in guessing at what was to come, do you think he would be very keen about such rewards? And would he envy and emulate those who were honoured by these prisoners and lorde d it among them, or that he would feel with Homer and greatly prefer while living on earth to be serf of another, a landless man, and endure anything rather than opine with them and live that life?"

"Yes." Said Glaucon. "I think he would choose to endure anything rather such a life."

"And consider this also." Said I. "If such a one should go down again and take his old place, would he not get his eyes full of darkness, thus suddenly coming out of the sunlight?"

"He would indeed."

"Now if he should be required to contend with these perpetual prisoners in 'evaluating' these shadows while his vision was still dim and before his eyes were accustomed to the dark — and this time for habituation would be very short — would he not provoke laughter, and it not be said of him that he had returned from his journey aloft with his eyes ruined, and that it was not worth while even to attempt the ascent? And if it were possible to lay hands on and kill the man who tried release them and lead them up, would they not kill him?"

"They certainly would." He said.

If you face the light
shadow will always
be behind you

Sukant Ratnakar

THE CRUCIBLE

SICK OF THE SYSTEM?

TIRED OF THE ESTABLISHMENT?

TIME FOR A REALITY CHECK

☾ THE CRUCIBLE

By way of introduction to Plato's Allegory of the Cave, the theme of this book, I bid you to a modern version of those prisoners born in chains, staring at shadows on a wall – called as they are to witness the eclipse of their world. I bid you to the Crucible, to a windowless room with a solitary chair. Please be seated. We are going to do a test. Come. Sit down. Make yourself at home, for we are making ready for a modern version of Plato's world of shadows. As you can see it's a simple experiment, a rudimentary affair, just a room with a chair facing a wall. Up and behind the chair there's a light near the ceiling ready to make your shadow on the wall, thus to perform the eclipse – an eclipse of a very particular and personal kind, for you are the object, heavenly body, in question, staring at a shadow, just like the prisoners in the Cave. For the moment the door to our room remains open and you can still see, but soon it will be as dark as can be, a necessary prerequisite to your birth in the Crucible, now the matrix of the Cave, as we turn your world upside down, inside out and the other way round.

For hitherto you have lived in a natural world, a world where objects cast shadows, or rather a world where, to be exact, objects intersected by light cast shadows. Like you, I don't see any reason to doubt this. It's self evident, we take it for granted, it's natural, perfectly normal, quite obvious. All this is about to change. Suppose someone told you the opposite. Suppose you are mistaken, you've got it all wrong, all inside out, upside down and the wrong way round. Suppose you have put the cart before the horse. Objects don't cast shadows. Rather, to the contrary, it's the other way round: it is the shadows that cast the objects, and we have been mistaken since day one: the real world is different in kind to what we know, and contrary to appearances, believe it or not, shadows cast objects. In other words, somebody has asked you to suspend reality and put your brain in reverse. If you don't mind we are going to go along with this, just for the sake of argument, for reasons which will soon become apparent.

Shadows cast objects.
Just three words.

Shadows > Cast > Objects.

It's a mahooosive mistake, no doubt, and daresay only a madman would believe it. Absurd as it may be though, we are going to entertain this notion. Might be as basic an error, as fundamental a mistake, as there has ever been, but we are going to go along with it, treating it like the holy grail. And, as you might expect, it has some interesting ramifications, for our simple premise is going to form the foundation of a strange new world: the Shadow World as we shall call it, a world where a shadow casts an object, is now the cause of the object – the object in question being you dear reader, presently disposed in the here and now. And the object in our test, this our Crucible, apparatus of understanding, could actually be any 'body' or 'thing' caught in the parenthesis of the shadow, but you will do perfectly for our experiment – as a living witness to the Eclipse.

For there nothing quite like the present, and will do it together, the Experiment – with a capital E, to distinguish it from the ordinary. As shall be Truth and Reality and all the rest. If you are sitting quietly at home with a book on your lap under a reading lamp, so much the better. You will see some sort of shadow, even your silhouette somewhere, and you can use it as an aide de memoir. The shadow may be on the wall, on the floor, on the bed or over the armchair or sofa, it doesn't really matter, but hopefully this will help us get into the spirit of the thing. Failing that one could hold finger over page as one reads, or even a pencil, as we take notes as the allegory unfolds.

Shadows cast objects.

You will note it's quite a simple experiment. A solitary chair, a wooden affair, in the middle of a darkened room with a light up and behind. Please come in and sit down, make yourself at home as you face the wall, and don't be distracted or alarmed as I close the door. Gently does it ... there it goes ... softly closed. Now it's as dark as night, quiet as can be. This is the chrysalis of our cave, the womb, and you have come here to be reborn: for the first thing you are going to see when we switch on the light is a shadow. Bathed in grey on the wall, it is all you will ever know, and it is going to take control, animate and haunt you wherever you go. Because whether you like it or not, in our new world:

Shadows cast objects.

Please get into the spirit of the thing. See it as a law or a commandment, as manna from heaven, or some sort of divine principle written on tablets of stone, laid down by the priests of some strange new religion. Give the religion and priests a name if it helps: The Grays, The Moonies or LunUz – or just Morons if you like. Or, if you are not religiously inclined, and this doesn't quite cut it, catch your imagination, imagine an alien civilization has invaded the earth, and this is the way they do things in a neighbouring galaxy. Call them Grays, feeble looking creatures, oozing with brains and huge cerebral heads: for they are not only more intelligent than us, but have an absolutely massive understanding of the universe. And considering our manifold ignorance, we had best take heed. Not that we can create the Shadow World for real. Admittedly it's beyond belief, quite unimaginable, beyond the ken of understanding, and would no doubt drive us up the wall to conceive of such a notion, let alone abide in it. But impossible as it is, absurd as it may be, we can but try and get into the spirit of the thing, just to get an idea of what it might be like. Hence our Experiment. It's incredibly simple and anybody can do it, thus to create certain conditions conducive to a certain outcome, the crux of this book.

As we said, the object, thing in question, heavenly body about to be eclipsed, dear reader, is you. For you are about to witness the eclipse for yourself, and seeing is believing in the world of shadows. Not that you can see your shadow as yet because it has not been made manifest: you are asleep in the chrysalis, waiting to be awakened from the womb, thus to become one in kind with those aliens, those Grays. The memory of your previous life outside the cave has gone, as has your ambient consciousness, window of your soul. For now you are about to be reborn into the world of shadow, and when this world comes to be as the light comes on, please remain motionless, staring out into the abyss, just like the prisoners in Plato's Cave. Play the game and gaze vacantly into the beyond of the beyond of the shadow on the wall. You are the Shadow. Treat it as a challenge. You are it and it is you. Anything other is an act of betrayal. Above all feel what it is like, get into the mode, the mood, as we take notes.

Ready? Let's make the switch, make the change, from one state to another, noting that there are no compromises here. It is either one or the other and there is no in-between. Once you were ON, now you are OFF – or rather … it's the other way round. Once you were dead ignorant, actually 'off', but now you have come to know the way, the truth and the life you are most righteously 'on', and the objective is quite simple: to become ONE with the Shadow. Are we ready? Nothing complicated, just you and your shadow. Quite simple. All we are going to do is change your mind.

Switch goes on:
Light > Object (ME) > Shadow.

Remember your head is constrained, your neck is fettered, so you are totally unaware of the light behind, just like Plato's prisoners, who, like you, were born in complete darkness. Click goes the switch, and here comes the Shadow … there you are … head and shoulders bathed in grey, now cameo to the wall.

Excellent. Time to introduce yourself.

Again, difficult as it may be, imagine you are the Shadow: you are It and It is You, meaning it is not only you, but the most true and just cause of you. Do your best, as if it's on the pain of death. Become a Gray or Moron, see it as set in stone, a holy precept, rite of passage, or just the comeuppance of fooling with Grays or communing with aliens. Suspend reality, quit reason, just stare at the wall and behold yourself as Shadow.

Note it happens all at once, for instantaneous and simultaneous, magnificent and manifold is the omniscience of the Shadow. Note the terrible distance, great divide between you and yourself as you are cast into the beyond of beyond of that infinite chasm, now the crux of your existence, crucible of your ego. Note you are no longer 'one', but utterly divided, cut in two, broken in heart, fractured of mind, as your world falls apart and credulity is stretched to its limits in all manners of thought and feeling. Your world has turned to dust, has been shattered like glass into fragments, the ramifications of which only a few we shall pick up, as we can but try to grasp, assimilate and comprehend, the consequences of such a monumental error, basic mistake.

The first of course pertains to the ontology of being: for now, to all intents and purposes, you are dead, or at least playing dead, as the shadow so conceived has nothing to do with reality as such, has nothing remotely to do with consciousness: you have been separated by it, metaphysically detached, so as to form a new awareness of yourself. The Grays, in their wisdom, might think differently, but we must admit, in this exchange nothing has been gained, no 'progress' here, or great 'advance', but rather a regression. Of yourself you are confounded, dumbfounded. Everything you are or have ever been, or ever will be, has been eclipsed, annihilated: your very vitality has gone, nothing remains except that truth which exists 'out there' in the Shadow cast over you. For we must admit, as witnesses to the Eclipse, we have been unspeakably reduced, impossibly belittled and impoverished.

Nothing wrong with the obvious. Let us state it, in case we missed it. Something's missing. Someone's pulled the plug, as down the hole it all goes ... our humanity, our sanity, ground of being, most basic instinct, touchstone of feeling, love, integrity ... colour and mystery, truth and beauty (and the discrimination of their opposites), heart, mind, reason, intellect, free will, spirit, life and soul ... down the plughole they go ... normality, common sense, our most natural ambient consciousness, window of soul ... out they go (not least our sense of space and time, now all dull and monotone) ... as down down they go .... unto the Shadow. And just to sum it up, let's give all this a grand title, something like 'The Rainbow of Consciousness', lest we forget, have not quite grasped as yet, what, after all, is at stake: all things bright and beautiful, all creatures great and small, all things wise and wonderful, the flower that opens, little bird that sings, all the hopes and dreams, rose of aspirations, glowing colours, reasons, imaginations, fruits of the human race, the whole meadow, moral and spiritual fabric, of Nature, Life and Soul, everybody everything you me and all, gathered everyday – in the Rainbow.

For the Lord God made them all.

But which God now? The Lord of the Rainbow or Lord of the Shadow? Which shall it be – the Rainbow State, or Shadow State – of our humanity? Because what exactly is the difference between me – the Rainbow – and the Shadow?

Rather big is it not? Mahoosive. For how great is the gap, that distance now! How deep! Vast! The sheer audacity of the act, duplicity of the wound, hurt to yourself – stung now – paralysed by the perplexity of it all, as bewildered and confused by the Shadow: a shadow that mocks you now, empty and pathetic creature that you are, stuck in these vacant premises, this prison of a cold and dark world. For what is this split, separation, great divide – call it what you like – that now exists between you and creation? By what just cause were you betrayed, taken of your sanity? What is this abyss unto which your soul has been cast? By what profanity? Wherefore this haunting image, this malevolence, this unthinking, this unfeeling and undoing of creation? Push the Logik as far it will go. Do the Experiment. Become unhinged. Make it your undoing. Come. Believe it. Come feel it, think it – be taken unto its dark duplicity. Like a malignant spirit, it just hangs there heavy in the air, all around and about, all outside in and inside out and everywhere. And yet closer than air it is, closer than breathing, as it invades your soul and takes over like a noxious substance that has brainwashed you – all courtesy of course to those Grays, those aliens.

Now that you have been brainwashed, knocked dropped dead unconscious and taken in, you cannot even begin to comprehend the magnificent act of self deception that is your ego – afraid as it is now of the selfsame shadow. For how did you come to be born in this alien universe, this vale of death, this stupendous world of anxiety, terror and despair? Were you born to die for a thought crime you never committed, of which you are truly innocent? Are you guilty as pronounced by the error, the biggest truism ever, the 'sin of the shadow' – a truth that cannot but be, a truth you cannot deny, cannot get out of, unto which you must surrender your days? For you may not betray the Shadow. As it holds you ransom to its object, so it binds you to its truth and chains you to its Logik. For now you taken prisoner, guilty of the construct. Something a fiend indeed. And considering the ramifications, is not this demonic phantom a personification of the devil himself – or rather yourself, the devil you have become, courtesy of the Crucible? For the more we advance, the more diabolical our world seems to get. Remember, it's extraordinarily simple, the Experiment of the Crucible:

For there is light. There is object. There is Shadow.
Object-Light = Shadow
Shadow = Object-Reality.
= TOTALITY =

And so the eclipse of the modern mind has been performed before your very eyes, courtesy of the Crucible. Case closed. Consigned to oblivion. Eternal damnation indeed. Truly fallen. Verily consigned. Unto hell, that void profound, infinite abyss, all thanks to the blinding science of those experts upstairs, who must by definition obscure the truth, clever and sophisticated as they are, coming as they do from afar, to take over the planet. For the Grays, in case you haven't noticed, are mission bound. In the profound depths of their intelligence, they have measured us up precisely for one purpose: the death of nature, annihilation of consciousness. The end of day on the earth. Totality indeed. Case closed. After all it's your funeral, felled like a tree so as to fall – ever so neatly and conveniently – into your shadow. Deeply complicated. Devilishly simple. For verily you have been taken in. Crucibled.

For as the light traverses object, so the body in question – you dear Soul – have been eclipsed of life and consciousness. Quite simple. For objects intersected by light cast shadows, do they not? Let's continue our object-light game, suspend all judgement and watch the outcome, following strictly as we do, the rules and methodology, laws and operational procedures of the Logik.

Note that you and the Shadow both move in perfect mimicry, (technically a one-to-one correspondence), as you are made to measure by the Logik. Now let's observe more closely the ramifications of this metaphysic as administered your soul. As you move so does the shadow, in an exact mimicry, as in a mirror darkly. You move, it moves. But this is breaking the rules. Lest we forget, it must move first: for it is the outside cause of you: not just of what you do, but of everything you are: for it is the Totality of what you think, say, feel and do. So we must be still, and wait for the Shadow to reveal itself. For the answer, the sum of our existence, is 'out there' after all, and as we stare out unto the Shadow, so we must sit still.

And wait.
How long?

You might ask.

Well, until the answer comes as you wait for the Shadow to act, be cause of all, the First Cause of your essence and existence, a cause that must be beyond in the beyond of the beyond of the abyss. The truth is 'out there' after all.

Just wait.

But how long?

Don't hold your breathe. Your guess is as good as mine. Bit like waiting for Godot.

Remember, never forget the thought crime. You of your own will or mind cannot make a consciousness choice: any movement, any action, any thought must come as if through some strange trick of mind to which you are blind: must be automated, mechanical, and when it comes, one must suppose, it must be experienced as such, as some sort of knee-jerk involuntary act. Note you must resist the entirely spontaneous and natural tendency of your mind to 'right itself' and recoil from the very absurdity of the notion of you-as-shadow. Broken and divided as you are, isolated and detached in a cold naked world of grey, you must reach a conclusion, make a decision, to have anything left of your sanity. Hard to do of course, considering your state. You of yourself have not the remotest idea of life or what it means, cannot feel or think or do anything. For when the revelation comes as it must, it must come from an act of involuntary will: must come from outside, must come from the shadow: must come about, seem to come about at least, through **Accident and Chance,** now presumably the premise of the Shadow World, a premise best imagined perhaps, as not just the way they do things in another galaxy, but another universe – a universe pulled inside out, upside down and the other way round at that. Very strange. Dark construct. For where exactly are we now, what happened in that switch, that change, that state? Are we on or off? Right or wrong? Up or down? Which way is it now?

The universe according to the Grays of course, who we must admit, just like the bondsmen in Plato's Cave, are busy up on the parapet holding all sorts of bodies and things – objects to the Light, objects that cast shadows on the Wall – thus to observe things in the 'object-light'. One must imagine, just as in Plato's Cave, we must hear lots of echoes, wall to wall communication – resounding success – as they bestow upon themselves all those honours and commendations and prizes as they make out the customary precedences, sequences, and co-existences in that 'one-to-one' correspondence to the Shadow. And the bigger the Shadow the better. For they have, what we presume must be, an absolutely massive understanding of the universe, and are busy seeking that correspondence by which we and all asunder are motioned and controlled by the Shadow so as come under the cause of the eclipse: the end of all consciousness. For Nature must be stripped of all value in order to get to the 'truth'. After all, you have no value in yourself, no mind, heart or soul, daresay freewill, for these have been exchanged: you have become a means to an end: the shadow of that State in which you find yourself. And the means will justify the end, because in the end there is no value, no witness to truth, appeal to reality as such, those virtues of Nature and Soul – the Rainbow.

Because what we may call life, daresay common reason and natural sense, is just refuse to the Grays, refuse to the process, a process which, incidentally, renders you incapable of make any decision for yourself in any situation, as 'the shadow knows best', has taken over control. In the mastery of the Matrix, in the vain attempt to measure and control everything, over their huge cerebral heads we must imagine the Grays have fuzzy O antennae, for everything must come under the radar for contact to be made, brought unto the touchstone of truth, brought unto the shadow, of what we must presume, is a kind of insect morality. Ghostly analogies, metaphors come forth, lost in the wordsworth of words that have no meaning, pertain to nothing, as love and honour, truth and integrity, heart, life and soul – out they go – lost in that abortive gulf, void profound, paradise lost unto which we have been bound – made to fit, made to measure, according to the construct, the underscore, the pleasure, given the oK as a good citizen of the Correctional Facility – a strange sort if 'Onited Kingdom' one must suppose. So motioned and controlled, so the door closes upon the green and pleasant land, window of your soul, and you surrender, acquiesce. All out of necessity no doubt, and in your own very best interest. It's just the way it is. The State your are in. Perfectly rational, quite reasonable, considering the premises. A foregone conclusion. As the Shadow makes over unto its own likeness whatsoever comes in touch with it, so everything becomes shades of grey, something of a performance, shady enterprise, a dark spectacle, as monkeys in a cage – or prisoners in the cave.

The subjects of this stupendous Experiment, have undoubtedly, (to put it mildly and water it down to intelligible terms), 'surrendered' the human spirit, not least their sanity, have sacrificed themselves to, been broken by the Shadow, and so must be put back together again like Humpty Dumpty, must pull themselves together in mass conformity to what we must presume are the new 'Laws of Nature' according to the Grays, busy as they are measuring up a new world order, truth correspondence. God forbid, one day our subjects might not even be able to apprehend any truth or natural reality, let alone change a light bulb, without some sort of 'shadow' overlooking them. To think of anything other, do anything for yourself, in the moment, right now, show a little common sense, open window of the soul, to them must be the great taboo. Measure up! So says the Logik of what we must presume is a rather Moronic System. Are the Grays prejudiced? Perish the thought.

We must presume, clever and sophisticated as they are, they would call it 'progress'. Progress being that which has been successfully rendered unto, brought under the Shadow. After all, nothing to them is true, unless it has 'object-reality', unless seen in the 'object-truth' of the 'object-light', thus to bestow the Shadow, a shadow they are no doubt proud of, and count it as yet another wonderful 'discovery' of truth and reality and the 'way things are' and how we all should act. Full of facts and figures, targets and statistics, and 'one-to-one' correspondences of theory and fact no doubt. Bit like living on the dark side of the moon we must presume. Or existing on that flat dimension called the Wall. All reflected in that dark mirror, that sliver of reality they call the 'world'. All courtesy of course to those aliens upstairs, those up on the parapet, those Grays, them what made Uz, the Universe and everythinK we are according to the LogiK, them whom no doubt – considering their claims to an absolutely mahoossive understanding of the universe – would be mighty offended if we called them rather shallow and superficial, let alone round the bend. Better perhaps, and more accurately we may call them 'Loonies', them that live on the Dark Side and dare not venture unto light. Presumably this would be the greatest taboo. For whatever is not seen in the 'object-light' cannot be true. And so they sentence their citizens to 'life' for a crime they know not of: for everybody must be downcast in the doom and gloom of the Cave, never to see the light of day.

Plato would be smitten, the allegory now proven: not just a world but planet entire, not to mention the universe, held captive to the darkness of the Cave. Lost souls indeed. No doubt, considering their condition, it would be difficult to convince them otherwise. Fraught with difficulties. To think that truth is just around the corner, that it's soooo obvious, sooo natural, that heaven (for heaven it must be, we must presume, compared to the shadow) is here – right now – seen in the glint of an eye. No way! Would say the bondsmen. That we dare presume. Ours would be another language, one they would find impossible understand. Mention the Rainbow and they would run like rabbits unto holes so as to go figure, correlate and observe and so measure how it cannot be any other, in any shape or form. To us what is as clear as light and day, they would have to make ever so complicated, difficult and obscure, just to impress us with their absolutely superior, most mahoosive understanding of the universe.

Oozing with brains, the Grays. Huge cerebral heads. Interesting place the Crucible. Simple Experiment. Basic choice. All through the blinding Logik of their Science. Mind you, nothing's really changed. After all it's just you sitting on a chair in front of a wall. Physically nothing has happened. You are the same as you ever were. Sat on a chair. Not that the Grays would ever get it. For we must admit, as far as they are concerned, it doesn't make a blind bit of difference. Or does it?

Watch carefully.
Those Grays up on the parapet, the 'bondsmen'
Watch the show.
Watch them hold an object to the light
So as to cast a Shadow
All in one-to-one correspondence
To the 'objective reality' of object-light.

After all its the pure reason, pure rules, pure conditions, the pure methodology, pure procedures, pure operational definitions – of a lonely planet caught in an eclipse. And it's so obvious, so basic, such a fundamental an error no Gray would ever get it. Or would we? Would it not eventually dawn upon us? Catch the light? That what goes around comes around, that the truth is as obvious as the lie. Nothing has changed. Or has it? Which universe do we live in now? And do we have to remain here, stuck with the Grays and their Moronic System?

Take a look around. Make yourself at home. If there's one thing going for the Grays, one thing they have got right, it's safety in numbers: so they can count you in. Don't worry, you are in good company, there's quite a few of Uz, masses of people in fact, anxious and confused, dumbfounded and perplexed – actually, if you think about it, a whole wide world, planet entire, staring into the abyss. Monkeys in a cage? The prisoners of the Cave? Lost Souls perhaps? Not that, fettered as we are, can we really be ourselves, even turn our heads, even see each other – dare question our existence – let alone the crime unto which we have been sentenced. But if we may be bold enough to take a glance up and behind, we may observe, albeit dimly, the bondsmen above. There they stand on the parapet, carrying objects and holding them up to the light, thus to create the shadows down below. Our prisoners are completely taken in by the shadow play, sit spellbound, robbed blind, downcast, born unto a life that is never theirs, a world of shadows, thus to witness the eclipse of their world.

For now the cave is the Matrix of the modern mind. Now the light of understanding is modern science, and the bondsmen are those men of science, those men upstairs up on the parapet, those masters of our understanding, makers of our reality, now the contemporary stage, world system, there to behold the objects and relations of our existence. Let the wall be grey and uniform, now the substance of our reality, given to the mass, measure and quantity of material and efficient causes. And so we may be worthy of the sin, that most original sin, the 'sin of the shadow', let us fashion our Cave upon a church, a sort of high temple, dedicated to the idolatry of science. For we are not even worthy of eating the crumbs under the table as the oblation is made, the object upheld, so as to perform the mass. And so we bow. Unto the holy communion, most blessed sacrament – of the shadow. And genuflect. In the logic of the triangle. For through it, with it, and in it we are made as of one Being, one State, one Substance. And let amen, the ancient of days, the genesis of our cave, bible of our defied science begin, through the aforementioned chance and accident, through an 'oops-y-daisy' sort of 'Big Bang'. For how other could the shadows come forth? For the answer is out there in the shadow you are staring, and seeing-is-believing in the world of shadows is it not?

Shadows cause objects.

For in the object-light – so they say – lies objective truth and reality as conferred by all, in the measure of those shadows on the Wall, shadows that are now cause to effect of the very object upheld. What an incredible parallel, exact correspondence, perfect correlation! Say our bondsmen. What a perfect cause, massive effect, one-to-one correspondence! For in the Shadow Kingdom, the shadows are not merely 'out there' as objective reality, but are the causal substance.

Shadows cause objects.

And how now, we may ask, have we come to live under the shadow of science? How did those men upstairs come to rule our hearts and souls – fettered as we are in the straight-jacket of materialism? What left us staring out upon this meaningless universe? Question is, can we bear it? Or must we mention some of its laws? Like seek and do not find. Chase the shadows of your mind. The truth is out there after all – staring you in the eye.

Measure up says the System. For all things bright and beautiful, all creatures great and small, all the reason and dreams of our humanity, are to be brought unto truth and reality. What is this place where ignorance passes for understanding, where everything has been turned upside down, inside out and the other way round? And in the doom and gloom are we destined to trip over the premises and stumble into the abyss?

Shadows cast Objects.

A simple error, catastrophic mistake? Are we beginning to see what's up – the state we are in, the shadow for what it is? For do we not put up object in front of our eyes to stop us seeing it? And have we not been made prisoners here – prisoners of our own device? Spellbound, robbed blind, born unto a life that is never ours, have we already passed away? For what happened to that heavenly body that once was you? What happened to the sun, your Soul, to Nature, the Rainbow of our consciousness?

Shadows cast objects.

We take it for granted. Don't question or doubt it. It's a self evident. Common sense. But suppose someone told you the opposite, that we have got our brains in reverse, the cart before the horse, that actually it's ...

the other way round.

Are we sitting comfortably?

Then we shall begin.

prelude

Eclipse of the light of heaven

eclipse of god

such indeed is the character

of the historic hour

which the world

is now passing.

MARTIN BUBER

OUR FATHER AUGUSTUS CAESAR,

WHO ART IN THESE THY SUBSTANTIAL

ASTRONOMICAL TELESCOPIC HEAVENS

HOLINESS TO THE NAME OR TITLE,

& REVERENCE TO THY SHADOW.

WILLIAM BLAKE

# PRELUDE

The hero of Plato's Cave meets a sorry end does he not? Perhaps he should have been a bit more careful before bearing witness to his experience, should have been a little more prepared, contemplated fate – at the least come up with some sort of reasoned defence. Might have saved him his life, demonstrated his brilliance, even proved his innocence, so as to leave something on record, some sort of legacy, testament of reason, lantern to humanity, before being condemned to death – just like Socrates, actually the true hero of the Cave, the real subject of Plato's Allegory, someone who accepted his fate happily, someone to whom Plato paid tribute in all his writings, so his legacy would never be forgotten and thus become the indelible mark of Western philosophy. Condemned by his fellow Athenians, death to the free thinking Socrates was no terrible tragedy, and he drank of the hemlock of his poison cup quite happily for the sake of philosophy. For the truth will set you free, a truth that, as far as he was concerned, opened the doors of perception to another realm, a higher divine world, inconceivably more beautiful than any mere mortality. For Socrates was quite happy to be rid of the body, itself a hindrance to the Soul, to take flight to a better place, the heaven of the Good, the Just and the Beautiful, a heaven to which Plato aspired in all his writings in commemoration of 'our friend, the best, wisest and most upright man I have ever known'.

For Socrates the human spirit, embodied in virtue, is sublime, but defiled by matter. We don't see the Beauty of the Good because we are stuck in materiality, and in the murky chaos of life our eyes develop cataracts that obscure the truth, and mistake appearance for reality in the illusory shadows of sense perception. And these shadows are not grey at all.

Far from it. They have the colour and form and pertain to the everyday world of the senses. As anyone acquainted with Plato's Allegory will quite rightly and quickly tell you, true Platonic shadows are quite unlike those shadows you were asked to endure in our test, the Crucible. Actually quite the reverse. In the Cave the shadows allude to the illusion and deception of the senses. They are not only in colour, but are a living and vibrant reality. By contrast in the Matrix, the modern equivalent of Plato's Cave, the shadows are really true, are easier to comprehend, for they are what they represent: they are obscure and pertain to matter as seen in the 'object-light' so as to form a blind spot, caused by that splinter in the eye, stick in the box, that 'body' or 'thing' held up, opaque to the eye. This blind spot is an occlusion of reality, is the umbra of the eclipse, cataract of mind – as obscured by the abstraction of material and efficient causes: reality now seen in the 'object-light', now the truth, now the 'objective reality', the underlying mechanism, the shadow behind the very sense reality of the phenomenal world.

In other words, as the Cave evolves into the Matrix, so the soul becomes substance, and the rainbow of mind becomes the shadow of matter – the first cause. Spirit, relating to infinite, transcendent, qualitative consciousness, is now secondary. Matter, relating to definitive logic and quantity, is now primary. One is exchanged and replaced, and thus eclipsed by the other, and this process, the defining characteristic of Western philosophy, takes hundreds of years. As truth is pursued, so it becomes the New Philosophy, New Science, as out of the frying pan of the Cave and into fire of the modern Matrix it goes. Because as the shadows change to material causes in search of truth, the reality behind the appearances, so the Allegory is on the rebound, is taken to a higher stage, shifted to another level in the attempt to resolve itself. Shafted, turned inside out, causally inverted, so it evolves into the final irony, most exact and exquisite parody of itself, as Plato's Cave becomes the dialectic of modern science, the matrix of our understanding, now no mere metaphor but mighty cause of the modern epoch. How the Cave became the Matrix in the crucible of our understanding, as witnessed in our introduction, is thus the crux of our story, theme of this book.

Let us extend that introduction by glancing behind the shoulders of present humanity, our neighbours in the Matrix, now no solitary cell but a vast expanse of people lost in pews extended in endless perspective. Here are the lost souls, those in the dark, those downcast, those who have been taken in. Come to your senses for a moment, wake up and look over your shoulder, up and behind your neighbours. Enshrouded in soft columns of light, you will see a parapet. Extending outwards like a shelf, it is the stage on which the objects of our world are held up to the light so as to cast the shadows down below. If you stretch your neck further and care to look even higher above our contemporary stage, you will see that our world shelf meets the base of a hill roughly shaped like pyramid. The pyramid tapers upwards, and stepped row upon row and cut into the hill are aisles upon which human figures are sat. These steps go back and up into history, rescinding into the past, each row an epoch of its own, each marking an evolutionary stage of humanity, each giving way down to the next, until we reach the bottom, now extended long and wide into the event horizon of our present world. Roll your eyes back from our living shores and you will see them lost in an ocean of thought, these figures, the masters of philosophy, makers of the Matrix. Cast as in stone from timeless rock, so they rescind upwards in the dark. Anaxagoras, Parmenides, Heraclitus, Pythagoras ... Plato, Aristotle ... Augustine, Aquinas, Kepler, Copernicus ... Galileo, Descartes, Newton ... Hobbes, Hume, Kant ... Marx, Darwin, Freud ... down to Russell, Einstein, Hawking ... they are all there, all present and correct, increasing in size and number as pyramid tapers down to the present.

Stretch your neck even further and follow the line of the pyramid up as far you can see, and you may observe a light in the darkness shining from above like a projector in a movie, and there at the top of the apex, right at the back and in the recesses, below a gentle wash of light, you may make out the faint outline of a man sitting in classic pose. It's Plato, the father of the Cave, his construct now no mere allegory or analogy, but rather a rational superstructure of such power and precision so as to at once liberate and enslave the world. The making of the Matrix, evolution of consciousness, stratification of the Cave, is a process that takes over a thousand years, and we are going to roll back the ages so as to comprehend the light shining in the darkness, the light that shafts our minds, that illuminates our consciousness, projection of our age.

And very Greek it is, this light shining from above, very Platonic, pertaining to geometry, a light linear logical, prism to understanding, a light to light the world, a light to unlock the secrets of the universe, shine upon our humanity and release us from the bonds of ignorance and darkness of a primitive past. Its true master is Pythagoras, of whom Plato is but heir, and the key is mathematics, unlocking the door to a new cosmos bequeathed by the Greeks. And the truth and way unto the promised land, this new dawn of consciousness, is proclaimed in the immortal words inscribed over the arch of Plato's famous Academy, now the formal entrance of our Cave.

Let no one ignorant of geometry enter here

Now this entry is calculating and exclusive, straight and narrow, and the door by which we enter is a pin point, a little peep-hole that defines the modern ego. In a moment, by way of gentle introduction, we are going to present one of the great Masters of the Matrix so we may find our bearings in the cave and put ourselves in the picture. But first let us remind ourselves, just once more, of the purpose of the exercise, as we extend the apparatus of the crucible to the world at large. For we may not be able to rationalize or explain it or put it into words, but maybe we can feel it, this shadow construct, right in our bones. It's like a splinter in the mind, a thorn in one's side, a background noise, a whisper in your ear that things are not quite right. Born in chains, fettered like the prisoners of Cave, bound in the straight-jacket of materialism, can we ever really be ourselves? For are we not prisoners here – prisoners of our own device? Grab on to what you can scramble for, hold on for dear life. Or rather sleep in the manger of the Matrix, so you may never wake up to the true ramifications of that conference of reality called 'the world' over and above on the parapet.

Which brings us back to those up above, and our first example. A few rows down from Plato you will see the 'Divine Engineers' of the Scientific Revolution: the likes of Galileo, Descartes and Newton. Ignore the figure in front with flailing arms. That's a recent contemporary. Stephen Hawking. Exalted indeed, hands up, all enthused, arms outstretched, waving his hands. For they never tire of getting up as close and high to the projector-light as possible, so as to make the shadow down below. Mind you, whilst we are here, it's a perfect example, demonstration, of one the most basic laws of the Matrix, and we might as well mention it in passing: the more you strive to get higher and closer to the projector light, thus to wave your godlike hands and play finger puppet, the greater the darkness, dimension of shadow, massive illusion, cast down below.

Or was it Richard Dawkins? Possibly Brian Cox? Or was it Popper, Skinner, even Imre Lakatos? Trouble is they are all so difficult to figure out.

We might tick off Hawking in a minute, you never know he might pipe down, but let's not get distracted and stick to our example, a real classic, let's take Galileo. Because you cannot understand Hawking – and indeed Stephen cannot even understand himself – until you understand Galileo, and you cannot understand Galileo without Plato, nor Plato, higher still, without Pythagoras. We will get to the top of the hill, apex of the pyramid later, but for the moment it won't do any harm to get to grips with Galileo, just as a little taster, little preview, to get gist of what's up, as we warm to ourselves to our task. Let's begin with one of his most celebrated statements.

> *Philosophy is written in this grand book – I mean the universe – which stands continually open to our gaze, but it cannot be understood unless one first learns to comprehend the language and interpret the characters on which it is written. It is written in the language of mathematics, and its characters are triangles, circles, and other geometrical figures, without which it is humanly impossible to understand a single world of it.*[1]

Here we may note in passing that this sort of attitude is going to get Galileo in trouble with the conventions of his time, in short the Catholic Church, whose language is the Word, or rather the Logos of the Bible. Not number, not mathematics: but words, and words, as we well know, describe the qualities of things. Numbers don't. And Galileo is having none if it. Rather he is establishing a new truth correspondence, new language and territory, he's minting a new coin, currency of thought and exchange value. If you want to buy into reality, there's only one thing that's going to get you to there, and that's mathematics. Without it we are entirely ignorant, because in truth reality is made up of those things which are measurable, things which can be positively defined by mathematics and are given to quantity: things which have solidity, shape, number, size and position. All these things are the same for all of us, for each and every observer. Rather than subjective, they are the objective properties of matter, because they really exist independent of the mind. They are really 'out there'.

Take a ball down an inclined plane. Work out its motion through abstract notions: derive a theorem thereof, that distance is proportional to the square of time, and it will apply to all such instances of motion, and such an experiment will allow us to predict and control all such instances. After all, it's the power of science – a science of which we are all too familiar – but behind this science and its claims to 'objectivity', to 'truth' and 'reality' and all the rest, lies something else at back of the mind – something you can feel, but can't quite define, something missing, obscured from view, pulled over your eyes. This 'something' is best viewed from the point of view of Galileo. It's only a little thing, and one might at first seem to be splitting hairs by taking issue with it. But within this little distinction, this little thing, so small you hardly notice it, lies the splinter that opens up the abyss. And Galileo here is most interesting, because he's right at the cusp of the whole process, right at the First Contact of the eclipse, figuring it all out, all the forces and motions, rolling balls down inclined planes and dropping cannon balls from ships, thus to derive not only the axioms thereof, but the metaphysics. Listen carefully as he talks of 'corporeal substance':

> *Now I say that when I conceive any material or corporeal substance, I immediately feel the need to think of it as bounded, as having this or that shape; as being large or small in relation to other things, and some specific place at any given time; as being in motion or at rest; as touching or not touching some other body; and as being one in number or few, or many. From these conditions I cannot separate such a substance by any stretch of the imagination.*

Nor we may add, can we. So far so good. Galileo is putting into words something we take for granted – that a thing is a thing by virtue of certain spatial attributes. Imagine any material or corporeal substance, anything, be it a ball or stone, cat or dog, planet or star – and you will find that it is 'bound' in some sense, bound in space and time, at motion or at rest, as one or many in number, in one place or other, of having some sort of shape or other. For to think of anything, conceive of any substance without some sort of space and dimension is impossible. Fair enough, what's the problem? Listen carefully, Galileo hasn't quite finished yet:

> *But that it must be white or red, bitter or sweet, noisy or silent, and of sweet or foul odour, my mind does not feel compelled to bring in as necessary accompaniments. Without the senses as our guides, reason or imagination unaided would probably never arrive at qualities such as these. Hence I think that tastes, odours, colours, and so on are no more than mere names so far as the object in which we place them is concerned, and that they reside only in the consciousness. Hence if the living creatures were removed, all these qualities would be wiped away and annihilated.*[2]

Watch carefully.

Galileo is up on the parapet.

Now watch him hold some 'thing' in the 'object-light'

do the eclipse,

As the light passes the object so its qualities

are wiped away, annihilated.

(prescient words indeed)

thus to leave the shadow, now the 'truth'

the 'objective reality' – the **cause**.

How now this 'objective reality' of the New Philosophy, New Science? Question being: what happened to the sun, the spirit of our humanity, soul of us, in that moment the crescent of the moon crossed over to darken our consciousness?

Not that we are questioning the efficacy of Galileo's method: it gives power, prediction and control. For mathematical correlations, as in a shadow, show a one-to-one correspondence with truth and reality, as in a ball down an inclined plane. For once you know the formula for motion, it applies to all instances of such motion: it's the sort of mathematics that put man on the moon. Rather we are questioning the premises: for Galileo must have his cause, must have his substance and his wall: and this substance is actually, as we shall see, an affirmation of the Greek philosophy of Atomism: that the real world is *different in kind* from the one we experience, the one we know. Rather it is a shadow, a moving geometry, the mechanism operating behind the appearances as a puppet upon a string. Now this is all very fine and practical for a physicist going about his business playing with balls, might even get us to the moon, question is though, is it true?

For even if we give Galileo the benefit of the doubt, and even accept in some shape or form that 'shadows cause objects', what can we say about phenomena that are not bodies, things that cannot be bound up, caught up and measured in the light, all those other Rainbow attributes we might care to mention in the spectrum of nature lost in the crucible of our understanding. Must they all go, must we surrender them, must they be annihilated? By way of example, let's take a natural object, a rose, and hold it up to the light. Note that as the light traverses the object all sensible qualities disappear, its colour, its scent, texture, form and beauty, as these are now made peripheral, at tangent to the intellect and senses: thus to leave substance. Question being, what is the difference between the rose and a shadow? Note that as you confer the shadow, the very properties you have impoverished it with have also impoverished yourself, because they are inseparable. However hard you try you cannot have one without the other. What you see now is a shadow of what once was, and what was once, is no longer: has been 'annihilated' in the process. Seeing 'through' it you have seen through yourself, and revealed yourself as you are: have become that to which you are blind. For you have discovered nothing real here, nothing sensible, no goodness or beauty, no spirit or divinity, no ineffable cause, no kinship of mind or soul, nothing remotely like the rose of consciousness – of yourself. Because ultimately there is no verisimilitude, no similarity, no resonance at all, to any real thought or feeling. For the rose of the world and world of the rose has been precluded by the very corporeal substance of Galileo, the matrix of the shadow, in our divided, disconnected world.

Are we beginning to get the picture, a glimpse of what's up? Galileo is just there at the cusp. Look behind and we can take anybody up there on the parapet more or less at random by way of demonstration. Let's watch the eclipse a bit later, fast forward it to a later phase, because this is a slow process, imperceptibly so. Let's take a few steps further down the cave to a more contemporary stage on the parapet. Let's listen to a celebrated physicist of the early twentieth century, Sir James Jeans, and take an excerpt from *The Mysterious Universe* published in 1931. Let's dip *Into Deep Waters*, the title of his final chapter. After a long foray into modern physics and the fathomless depths of the quantum world, he concludes:

> We discover that the universe shews evidence of a designing or controlling power that has something in common with our own individual mind – not so far as we have discovered, emotion, morality, or aesthetic appreciation, but the tendency to think in the way which, for the want of a better word, we describe as mathematical ... we cannot claim to have discerned more than a faint glimmer of light at best; perhaps it was wholly illusory, for certainly we had to strain our eyes to see anything at all.[3]

No change there then. Jolly good shew. Straining our eyes indeed. Uzzz. Fast asleep.

Let's take a step down further to our contemporary stage to see if anybody has woken up. Quick dive. Dip into Amazon. See who wriggles out top of the algorithm. How about that brain of Britain – Brian Cox:

> What is meaning? I don't know, except that the universe and every pointless speck inside it means something to me. I am astonished by the existence of a single atom, and find my civilisation to be an outrageous imprint on reality. I don't understand it. Nobody does, but it makes me smile. This book asks questions about our origins, our destiny, and our place in the universe. We have no right to expect answers; we have no right to even ask. But ask and wonder we do. *Human Universe* is first and foremost a love letter to humanity ...[4]

A *human* universe? Well ... not quite. Pointless specks? Count me in. Outrageous imprint? Sh ... got that one too. Origins. Mmm ... can't quite make up my mind. Destiny? That's easy. Durr ... kind a dark. Place in the universe? Yep, got that one too – I'm on it right now, reading the love letter. Tough one that. Hm ... now what happens when something is divested of essence, its qualities wiped out, taken away, is well and truly digested ... now what ooozes out at the other end of this humongous 'shadow' process, what comes out? Errr ... I'm on it now, can feel it coming .. straining on my seat in the cubicle, tight in suspense upon my throne. And you know what, you're right, in a funny involuntary way it does kinda make you smile. Can't bear the suspense, wait for the day the answer comes, the day we are all truly astonished – by all the Russells n Dawkins n Hawkings n Grays n Morons – the Toiletarians of our world. And no wonder no one understands it, it's crap. ESSENTIALLY crap.

Sorry, couldn't constrain myself. Been waiting to drop that one for a long time. Irony and profanity and outrageous imprints aside, can we not but **smell** the hypocrisy? And it's not just Brian who's been pigeoned, who's the butt of the joke, but aren't we all? For are we not all locked in our cubicle, the crucible? And it stinks. And don't we know it, doesn't it half echo in the wall to wall communication of a world, that really is an outrageous imprint – full of _IT. Profanity indeed. Astonished. Because none of us, so it seems, have an atom of sense, remotest idea of reality. And that atom, the 'existence of a single atom' that astonishes Brian, is what splits the difference. And the distance is not far. It's tiny, right on the cusp of Galileo, the First Contact of the eclipse. Because is it not about time we got the metaphysic over and done with, all this out of our system? Huge relief for all concerned. Long time coming. Feel so much better. Then we can all make cupcakes.

As for Stephen. It's a classic. Bestseller. It's 1988 and he's in search of the answer to the universe and the final end of physics through a unified theory of everything in *A Brief History of Time*. Once we have found the answer, he suggests in conclusion of his book:

> The we shall all, philosophers, scientists, and just ordinary people, be able to take part in the discussion of why it is that we and the universe exist. If we find the answer to that, it would be the ultimate triumph of human reason – for then we would know the mind of God.

What?

in a   shadow?

Deus ex Machina

GENESIS

In the beginning
God created the heaven and the earth.
And the earth was without form, and void; and darkness was upon the face of the deep.
And the Spirit of God moved upon the face of the waters.And God said, Let there be light: and there was light.
And God saw the light, that it was good: and God divided the light from the darkness.
And God called the light Day, and the darkness he called Night. And the evening and the morning were the first day.
And God said, Let there be a firmament in the midst of the waters, and let it divide the waters from the waters.
And God made the firmament, and divided the waters which were under the firmament from the waters which were above the firmament: and it was so.
And God called the firmament Heaven. And the evening and the morning were the second day.
And God said, Let the waters under the heaven be gathered together unto one place, and let the dry land appear: and it was so.
And God called the dry land Earth; and the gathering together of the waters called he Seas: and God saw that it was good.

The Timaeus

When the father and creator saw the creature which he had made moving and living, the created image of eternal gods, he rejoiced, and in his joy determined to make the copy still more like the original: and as this was eternal, he sought to make the universe eternal, so far as it might be. Now the nature of the ideal being was everlasting, but to bestow this attribute in its fullness upon a creature was impossible. Wherefore he resolved to have a moving image of eternity, and when he set in order the heaven, he made his image eternal but moving, according to number, while eternity itself rests in unity; and this image we call Time … Time and the heavens came into existence at the same instant. God made the sun so that animals could learn arithmetic – without the succession of days or nights, one supposes, we should not have thought of numbers. The sight of day and night, months and years, has created knowledge of number and given us the conception of time, hence came philosophy. This is the greatest boon we owe to sight.

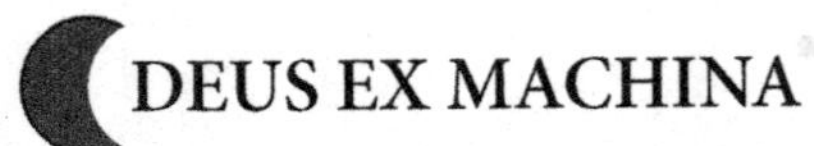 DEUS EX MACHINA

The Logic of the Logos

Did Plato's Allegory remind us of someone else? Not Socrates, but some other prophet, hero, daresay messiah? Stay a moment, who comes to mind? Plato writes the script and sets the stage and so comes Christ – the cave now no mere allegory but the gospel truth, making a remarkable prelude to the New Testament story of he who awakes and takes the stage, speaks in parables, performs wonders and miracles, so as to dispel the shadows of the cave. And he has the temerity to call himself God and the cavemen children, so as to receive not so much the poison cup but crucifixion – from which he descends to give the final proof in the triumph of the Resurrection. For above the Cross stands the immanence and transcendence of the Rainbow: Jesus Christ.

Both Jesus and Socrates proclaim love, one of the mind, the other of the heart. One was condemned by Athenians, the other by Pharisees. Neither wrote anything down, left the record for their disciples, who wrote in the form of question and answer, parables and dialogues, out of which emerged the two great bookends of Western civilization, The New Testament and The Republic. And in between lies the story of the West, those chapters of history and natural philosophy that built Christendom and the City of God, modern civilization and the citadel of science. Both Christ and Socrates were called heretics, were made scapegoats, said virtue was the true path, said the body counted for nothing, appearances didn't matter, and whereas Socrates went about the marketplace asking questions about virtue, Christ just turned the tables. For the Christian metaphysic was more radical than any Greek or Socrates, for whom a table was only something to write on, was a mere scribbler compared to Christ, who could overturn the laws of nature, heal the blind, calm the storm, even save a wedding out of wine. For he was divine, and in act and fact knew a bit more about the relations of spirit and substance, mind over matter, more than any scribbler or philosopher, now or ever since. And there-in lies the rub of Christianity and the turning of the tables of heaven and earth and that ever so intimate relation with God: there-in lies the difference between the Platonic and Christian world.

These two sides, Socratic reason and the Christian spirit, are going marry to make the matrix of the cave, thus to conceive a child born of the love affair between Christian faith and Greek reason. This affair eventually ends in the divorce of the church from science, and the Greek from Christian. The relationship, successful as it was, was doomed from the start. For, on the one side was the Logic, on the other the Logos, and it was the 'Logic of the Logos' (to coin a term), the conceptual marriage of Greek and Christian ideas. that formed the cave of Christendom, the matrix of the modern mind.

By way of demonstration, we have two creation stories, *Genesis* and *The Timaeus*, one of the Bible, the other of the Greeks, one we are familiar with, the other perhaps not. Apart from those Creationists who still ascribe to a literal interpretation, Genesis today seems a fancifully naïve explanation of the beginnings of the world. We know so much better now. Or do we? Do we know that our modern vision of nature is indirectly derived from this account, and that it may possibly be, in it's own way, just as primitive and naïve? That is to say we read 'objectivity' into our science just as the ancients read 'objectivity' into the Bible – as to the truth and way the world was created and how things really are.

The difference between our two accounts is that of language. One story is written in a language which pertains to qualities, the other to quantities: one is Genesis, the other The Timaeus, one pertains to the Word, the other to Number, one is Creationist, the other Mechanist, and both are equally and distinctively made by the same 'Father'. For unlike Greek or Latin, or just plain English, mathematics provides a far more cool and precise interpretation of how God's will is realised upon earth: through a concealed mechanism, and it is by no mere coincidence that the originators of the New Philosophy during the Scientific Revolution saw God as a 'Divine Engineer' presiding over a perfect clockwork mechanism.

In modern times this curious coincidence of faith and reason has been conveniently forgotten and ignored, for the whole makings of science and its faith in objectivity have crucial religious assumptions, taken for granted today, but vital to its claims to 'objective' truth. If we look back into the recesses our Cave we may note admittedly dimly and briefly, that it grew out of a Biblical past, together with its dualistic preconceptions.

For the God of the Bible is separate and apart, so separate as to become an object of thought, a universal first cause, possessing an almighty power which determines the world. And He is not pagan or superstitious, given to paganism, images or idols, but rather He is a God of law, who fashioned the world from the 'outside', as a potter from clay. Omniscient, omnipotent, of definite being, so God possesses an almighty and absolute power over Nature and nothing can contradict his Will. For He is perpetually external, overseeing creation, 'up' and 'out there' as an object of thought, judging and weighing up the world, determining the course of history through divine laws. Implicit in this account is a dualism – a separation of God from object, of spirit from nature, mind from matter – that forms the primitive genesis of science. Animism, paganism and idolatry are outlawed and forbidden. For the God of Israel is a mighty God and will not tolerate pale reflections, will not lower himself to hollow appearances and graven images, is remote, high, mighty, a God who becomes not so much spirit but substance: an object of Greek thought as fashioned through the works of Descartes and Aquinas. Logical relations now establish how God made and sustains the world so as to become the Divine Engineer of creation. As Christianity is cultivated by Greek thought, God is made over to a mathematical creator in a new synthesis of religion and reason, now the perfect Greek answer to creation – The Timaeus, the rational explanation of Genesis, through a concealed mechanism operating behind the world of appearances. The Logos of the Bible, the story of how God made the world, was realized through the Logic of Science. As the Logos became Logic, so God became Mechanical in that process called the Scientific Revolution.

Not that modern science any longer bears any explicit resemblance to its religious origins and assumptions, for we are no longer aware of the cave and blissfully ignorant of the premises. Only be saying, as did the likes of Descartes and Newton, that some other mind, some other separate being created a world from the 'outside', could the New Science merely be said to be going about its duty of 'discovering' the truth in the deified mirror of mathematical relations.

With hindsight we may note the tragedy born of this idolatry, now a foregone conclusion: the Jesus of the cave is exorcised of love by the cross of science, the irony being that the nascent church – bequeathed with the problem of how Christ-the-Logos was related to the Father, how the human is related to the divine, how matter is related to spirit – should have come up with an answer that was to become the apostasy of spirituality. For God was now the Logic of a dead abstraction born of definitive geometry, was a Greek Demiurge beholden to the Summa Theologica. This process, the Logic of the Logos, is the work of centuries of thought culminating in that immaculate conception called The Scientific Revolution, born as it was of the marriage of Greek and Christian ideas, now the 'Deus ex Machina', world machine, the final solution, as conceived of Genesis and The Timaeus.

In one account the world was made from the Word, in the other from Number. One from the Logos, the other from the Logic. Both in the image of the Father, a divine creator. They are two sides of the same coin: each explains the other. How did Genesis happen? How did the world begin? Through numbers and geometry, through natural science. The Greeks will tell you that. Move that geometry, give it a First Cause and Motion, and so we derive a far more cool and precise interpretation of exactly how God's will is realized on earth: through a divine clockwork, concealed mechanism.

Heads or tails. Which wins? Science or religion? The heart or head? Genesis or The Timaeus? We know the upsum – it's the face of science. But at the end of the day which truth wins, which shall be timeless? Now which will it be, the Logic or the Logos? Or are they really deceptive truths – doubled faced realities – like the Greek god Janus? We know that today the odds are against religion, for in our cave, The Timaeus is the safe bet, real deal, true currency in the daily dealings of our lives. Genesis, except for a few creationists, has lost its worth, foregone by history, is no longer sovereign, has no exchange value.

Let's see how we got here, watch how it all began. Let's behold the coin before it ever was fashioned and divided into two, minted and so impressed upon our minds. Let's toss it into the dazzling sun and watch it sink into the ocean of the Greeks, as we return to the entrance of the cave to behold another dawn, one that went before. Let us meet in the Golden Age, to ponder the genius of those stars that guided and brought us here, retracing our steps along the shore.

The Greeks

☾ THE GREEKS

**The Milesians • Heraclitus
Parmenides • The atomists
Pythagoras • Plato • Aristotle**

•

Every age has its cave with a chandelier on the ceiling casting light unto the darkness below. Imagine the crystal chandelier as a Newtonian prism. On the one side is the world of colour and consciousness, the spectrum of nature, the life of the spirit, now the Rainbow, and on the other is a narrow beam of white light. Imagine this light as reason, now a straight line, pertaining to abstract logic: as the rainbow passes through the prism, the spectrum of experience is rarefied into correlative logical principles via the straight and narrow path of light. Now this abstraction of nature into the linear logic, this reduction of phenomena to intelligible principles, is the aperture of our cave, door to modern perception, and it begins with a Greek fascination for geometry, now a dazzling crystal in the recesses of the cave. This chandelier, this prism, crystal logic, is all focussed and precise, giving power and insight like never before, thus to fashion a new dawn and evolution of human consciousness through that great epoch known as the Golden Age of the Greeks.

Just in as the Christian picture, in Greek philosophy there is an old and new testament, those who come before Socrates: the pre-Socratics: Thales, Anaximenes, Anaximander, Heraclitus, Pythagoras and Parmenides – and those who come after Socrates, the post-Socratics: Plato and Aristotle, the Epicureans, Stoics and Cynics alike, together with that great synthesis called neo-platonism, of which the enigmatic Plotinus was founder. Just as the Bible was a unique religious picture so Greek thought was a unique – and unprecedented, philosophical outlook. Despite dwelling in a pagan world this outlook was different from anything that had come before. To the Greeks the Kosmos was a sort of playground, and philosophy an Olympian sport. Not only did they dare ask the big questions, they had the confidence and courage to answer them. For the Greek was an exceptionally inquisitive child, and this is what distinguishes him from the rest of creation. His eyes were not only open wide but furrowed, and while he marvelled at the mystery of creation he came up with a reason all by himself, without recourse to anyone else. Inventing a cosmos from scratch was something he felt free to do, for his itching curiosity was most peculiar, so peculiar as to be logical and systematic, forming a primitive science of nature.

And this science emerged first and foremost as a fascination with geometry. As the likes of Galileo and Descartes were to attest, the underlying rationality of the universe is actualised by the timeless purity of geometrical relations. These relations not only embodied Greek art, later to be revived during the Renaissance, but the modern cosmos to come. For the world was not just geometry, but a moving geometry, the machine that was to become the clockwork universe of Newton, a machine that was as yet quite alien to the mind of the Greek, who breathed a different pneuma, all reposed under a divine, living sun. For the Greek mind was spiritual, still partially 'pagan', was different in kind: their 'stand under', the world that upheld them, their substantia, (sub: under, stare: stand), was the physis, literally the spirit of nature, essence of things. Nature was not dead or inanimate, but rather a living spirit, a beating Rhythmos, an organism. Gods and spirits were external representations of what was sensed inside: for what was outside was felt within through an emotive connection with the cosmos as a whole. As above, so below. It was a human world in every sense, for they, the supernal gods, were like unto us, were emotional, got angry. For is not an earthquake the phenomenal act of a mighty Spirit? Certainly is if you are in the middle of it, an act of god, a superhuman earth shaker Poseidon of a god, and to the Greeks all natural phenomena were animated by gods. Like Aphrodite, the goddess of Love And Beauty, not angry like Poseidon, no curly hair or white beard, not out for retribution: but then that's not Her role, her function. Want to produce a beautiful work of art? Then turn to her for inspiration, pray and meditate upon her object, see her poetry catch the light, inform all things. For the Aphrodite of our Beauty cannot be an 'it', a mere thing, for She is the timeless apple of our eye that transcends space and time, as do other gods like Ares, the god of war that ravages civilizations. We see her now and again, Aphrodite, in the mystery and beauty of life, just as we are want to ponder our own loved One. What makes her Beauty, and where does it come from? And why is she subject to eruptions and quakes? Ask Poseidon, for in Greek mythology he is not only God of Quakes but also those fathomless depths of the Aphrodite of our Sea.

•

THE MILESIANS

And from these depths comes the greatest intellectual earthquake ever to shake the earth, and it begins around 600 BC in the Aegean sea next to that great European swimming pool, the Mediterranean, a world not only rich in trade but flourishing with ideas, as science encounters its baptism by water. For all it mattered the first element could have been anything, could have been bread or butter, or chalk or cheese (or even treacle, one wistful suggestion from a Cambridge professor), for what matters is that this 'water' is the first sort of thing, first notion of substance, first idea of natural causation as it evaporates, condenses, and encompasses all things.

Such was the opinion of Thales, the first of the Milesian

philosophers, who duly recorded the first element of natural science, the first substance underlying a world of change: for behind the seeming chaos of nature there exists an underlying form, a persistent stuff conserved throughout all. Not that this elementalism was unusual in the ancient world. The Chinese had it. What was unusual was how it developed into the abstract thought characteristic of modern science, as the elemental stream accumulates into a river and meanders through the eddies of Heraclitus, the still pools of Parmenides and rocks of the Atomists, before it reaches the great ocean of Plato and Aristotle, only to be dried up and washed away by the Dark Ages.

From Thales to Anaximenes we go from water to air, now the lifesoul, as modified by rarefaction and condensation as it thickens and thins into the elements of earth, water, wind and fire. And this element is rarefied again by Anaximander, Anaximenes namesake, whose air becomes the *apeiron*, disappearing from the senses altogether so as to form an indefinite substance, elixir. Infinite, ageless, hidden and unseen, the *apeiron* manifests itself through *dike* as a juxtaposition of elements perpetually rebalanced by nature. *Dike* is Greek karma: what goes around comes around, a dynamic law, and nature as well as human society is subject to it: for anything that disturbs the balance of nature does not last long, and must be redressed in time, or in the words of Anaximander. 'Into that from which things take their rise they pass away once more, as is ordained, for they make reparation and satisfaction to one another for their injustice according to the order of time.'

The elixir is rarefied higher unto abstraction with coming of the Nous — the Soul of the universe, governed by reason, as Xenophanes would declare. Implicit is a scepticism of the senses in favour of logic, as abstract ideas now manifest the world, a world now in motion or at rest, perpetually changing or standing still: two conclusions reached from opposite poles, one through Parmenides and a love of pure logic, the other through Heraclitus and his contempt of it — a polarity Heraclitus himself would have revelled in: for life is like a stream, all in flux, full of eddies and pools, riddles and paradoxes, tensions and opposites, which the mind can never fathom through abstract logic.

•

HERACLITUS

For Heraclitus the world is an everliving fire. And fire is spirit, not a thing, is a manifestation of divine energy, a chain reaction as opposites coincide, all fused together into a whole whose principle is strife.

We must know that war is common to all and strife is justice ... all things comes into being and pass away through strife.[1]

Again dike is the crux: all things come to be and pass away as it redresses the polarity of nature.

Men do not know how what is at variance agrees with itself, it is the attunement of opposite tensions, like that of a bow and the lyre.[2]

For the world and its objects of consciousness are neither here nor there, are neither this nor that, rather they are like a living stream, an unbroken wholeness:

You cannot step into the same river twice, for fresh waters are ever flowing upon you.[3]

The flowing of the river is the flow of consciousness, never still, always moving, in a dynamic flux of tensions and opposites, like the strings on a lyre: and only in the attunement of opposites can harmony and disharmony exist, according and discording, forming and unforming, as every moment is created anew by the Logos, the formative principle, the Reason behind the play of the universe, based on one fundamental principle, the union of opposites: opposites that work out of pure necessity. For in the world of opposites you cannot have one without the other, love without hate, or pain without pleasure, for they go hand in hand. To be wise is to transcend the appearances, not to become drawn into paradoxes, not to get stuck in rocks of abstraction, or caught up in the eddies of the moment. To be One is to know the Logos and sit by the river at home in the ebb and flow of life, there to be with nature just like the Taoists of ancient China. For Heraclitus is of the East, and he sits more at home in that most iconoclast citadel of all, a Zen Monastery, with his paradoxes and koans.

Time is a child playing draughts, the kingdom is a child's.

The way up and the way down are the same.

How can you hide from what never goes away? [4]

Like the Tao, Heraclitus is all up and down, is neither here nor there, is the sound of silence, the warring of love, of light and darkness, the bifurcation of pain and pleasure, all mixed up together in a whirling cosmos revolving around an everlasting fire.

It was ever, is now, and ever shall be an ever-living fire, with measures kindling and measures going out. [5]

For matter, from the lowliest stone to the celestial star, may appear to be dead and inert, but actually it's strung upon a lyre, tautened tight as on a bowstring, full of infinite energy and potential. Lifeless, dead and inanimate as the stone may appear to be, we cannot grasp the great Heraclitean bow that upholds it, stretching like a rainbow across creation, holding it all together, one positive pole and the other negative, in a world which, as many cultures attest, is literally strung, or rather sung, into existence.

•

PARMENIDES

If Heraclitus was the Lao Tzu of the Greeks then Parmenides was the Buddha. If you have ever felt a timeless now, the eternal stillness and peace of nature, or just want to logic yourself to nirvana then Parmenides will take you there. For the stream of life may appear to be moving but actually it's standing still, immobile in the supreme bliss of the 'IS', now and forever:

It is unborn and imperishable, whole unique, immovable and without end. It was not in the past, nor shall it be, since it now is, altogether, one and continuous.

So says Parmenides through a pristine logic that characterises the Greek from the 'irrational sensation' of the pagan:

In my opinion we must first of all make the following distinction: what is it that always is and has no becoming and what on the other hand becomes continually and never is? The one comprehensible by the mind with reasoning, the other conjectured by opinion, with irrational sensation, coming to be and passing away, but never really being. [6]

For Parmenides has a simple proposition: you cannot conceive of what is not. Let's pause for a moment in the Parmenidean crucible. Try to conceive of nothing, of simply not being: of being that which is not. Try to conceive the inconceivable. Impossible?

For you cannot conceive of what is not.

Now the premise of Parmenides, for he is the first supreme Greek, abstract thinker, philosophical Hamlet. 'To be or not to be' is the question. And 'not to be' cannot be the answer, is

most inconceivably out of the question. The end result is a uniform, timeless and unchanging substance behind the manifest
world of appearance, something that always IS and cannot but
be, now the One of which the world is but an illusion. Talking
of Parmenides, a commonsensical Aristotle says:

*Although these opinions appear to follow logically in a dialec-
tical discussion, yet to believe them seems next door to madness
when one considers the facts. For indeed no lunatic seems to be
so far out of his senses as to suppose that fire and ice are 'one'; it
is only between what is right and what seems right from habit
that some people are mad enough to see no difference.* [7]

If for Heraclitus everything was change, then for Parmenides change was impossible, logically impossible, quite inconceivable. Because that 'something' which you are thinking
of now really does exist, and can be nothing 'other' than what
it is. And it does not only exist now but always lives in the IS
of Parmenidean logic:

*What is cannot have come into being. If it did, it came either
from what is or what is not. But it did not come from what
is, since if it is existent, it did not come to be, but already is;
nor from what is not, for the non existent cannot generate
anything.* [8]

Nothing can move, nothing can change, for it cannot 'be'
and 'not be' at the same time. Either it is or it is not. Just as
in the binary logic of the computer there is no in between.
Something cannot come from nothing. It's a contradiction in
terms. And the computational logic of Parmenides says that
we can only know what IS: to know what is not implies self
contradiction.

Everything simply is. Can be nothing other what it is.
We cannot not be. Everything simply is and can be nothing
other than what it is, for to 'change' involves a self contradic-

tion, because one thing cannot change in the 'other', become something else, it cannot both be and not be at the same time in the same place. Either it is what is, or it is not: it cannot become anything other than what it is, do a disappearing act and miraculously become something else. Nothing changes. All movement is an illusion, is of appearance only, because it implies change, not from one aspect to another, but from one place to another. Something cannot both be and not be in one place at the same time, for this what change is: a dynamic process involving these aspects.

All is One: one unchanging timeless substance through which the phenomenal world manifests itself, and this Greek substance forms the primitive origins of modern science because it is perfectly logical, computational and non-contradictory: seeing the world in terms of a 'one-to-one' correspondence of ideas and reality: for everything correlates to the one unchanging substance. And this One is bound by a pure logic of which the sensible world is but an appearance every changing and illusory and seeming forms. Water may be cold, the sky may be blue, and roses may be red, but these are qualitative statements and mere associates, metaphors and analogies that not only make up the appearances but obscure the truth.

For how can water be cold? Or roses red? The sky blue? Water after all, can be nothing other than what it is, cannot be anything else than what it actually is: water. Water cannot be cold any more than roses can be red. They are one thing or the other. And even if we were pushed to define what we are talking about, and looked it up in the dictionary, we would always find another qualitative analogy, dictionary term, heuristic definition. All we can say at the end of the day, is that water is Water, a rose is a Rose and sky is Sky.

In other words they are what they are. But then they vanish, they disappear, have no boundaries, no attributes or distinctions unto each other. They, we, and everything IS. Being nothing other than what they are, they are One, for whatever boundaries we put upon them in the garden of our minds are false constructs of thought born of the categorising nature of language: thus they are an imposition upon reality, which truly is transcendent and One. For we are all one substance, one totality, one whole, the IS.

Parmenides called a spade a spade and made a metaphysical issue of it. Nothing can be anything else, other than what it is, cannot change into something else, because all things are formed by one unchanging substance, the frame, the canvass, which is necessary for anything to exist at all, be perceived of as whole, as time and change play out their days in the grand illusion of the 'world', now the Parmenidean stage, 'maya' of thought. And this One is the wonder of the Greek mind in its perfect purity, symmetry of form. For spade will be a spade. And water is water as a rose is a rose, and nothing other. For, after all (if we dare dwell long enough on it) how can water really BE (cold), how can it be different and 'other', yet one and the same?

Parmenides sets the problem, and the Greek mind finds itself at an impasse, for knowing anything has become something of an impossibility, and the world of the senses and reason are far apart. For how can we know the One? We cannot see it, touch it or hear it. After all it's an abstraction. Something you cannot see. We know 'oneness' as a feeling. But Parmenides is not exactly being romantic, but rather abstract, so abstract that only one mode of thought seems to represent his persona, working as he does up to a mental level commensurate with the abstractions of mathematics, where one predicate, one group, one set is exactly the same as the other.

Water is Water as W=W just as 1=1 and 1+1=2.

They are ontologically the same, similar groupsets, correspondences and integers to the One. In terms of logic these are called analytical statements of contradiction, are object quantities because they are logical necessities. But is the Creator,

God himself bound by this logic – this Logos – the logic of the Greek? Because if he is, he is no longer free, is no longer 'God', but rather what the Greeks would call a Demiurge, a sort of deified draughtsman, whose works are subject to the rules of logic and mathematics. Might have built the cosmos, but had to use the Logos in the process, had no choice, was actually Logic himself, or rather itself, which makes him no longer God but something rather more distant and remote, abstract: a mere automaton. Yet Perfect. As perfect as Logic. He is incapable of self contradiction, especially analytical statements of non contradiction. And these are 'necessary truths'. To deny them is to deny the very nature of reason, they can be nothing other than what they are as expressed by the timeless purity of mathematics, the substance of Logic. To say 'water is cold', roses are red or I love you doesn't really count, doesn't cut it. But you can bring flowers: as long as flowers are flowers and only flowers and nothing else. That's fine, as long as they are uniform. But you can't say they are violet or blue, or they smell wonderful. Doesn't quite cut it. Not strictly true. For this is illogical, implies a self contradiction: for whether they be beautiful or ugly is irrelevant, is to express a truth that is neither digitally correct nor mathematically commensurate.

•

THE ATOMISTS

If all we touch, see, hear and feel is an illusion, and nothing every really changes but has its origins in the same thing, permanent substance, then how is this One related to the manifest world of appearance? Where's the causal link? How does it happen? The perfect – or almost perfect Greek answer, was the Atom. A simplex thing, an irreducible element, imperceptible to the senses, hard and indestructible: it was the exactly the sort of brick a Demiurge would use. For it behaved itself, was a thing, a pure quantity that obeyed the rules according to number, size, shape and position, moving about in kaleidoscopic certainty. The single uncompound unit of existence it was the One mirrored and manifested in the Many. In short, it was a stroke of genius: the first ever mathematical object of thought. A number object, something operating behind the world of appearances to which all and sundry were puppets. Not water. Not air, not fire, but a thing, a hard and fast thing. For in the beginning there was only a primordial chaos of atoms. Gradually over aeons of time these atoms came together through accident and chance to form the consistent pattern of the cosmos, their microscopic interactions determining the phenomena of the macroscopic world we see.

Greek Atomism lies dormant for over a thousand years before it germinates under the right conditions of the Scientific Revolution. Cultivated by notions of mass and the separation of mind and matter, Atomism contains the crucial premise of the New Philosophy: that the objects and relations of the real world are different in kind from the world know by the senses, different in kind because they are quantitative and mathematical. These are to be the 'primary qualities' of Galileo. They come first and necessarily so, for they are the micro-mechanism operating behind the world of appearances, those 'secondary qualities' of macro-sense and experience. It is primary qualities (which really are a misnomer, because they are not qualities at all), or rather primary quantities, which make up the secondary world of experience, their microscopic interactions causing macroscopic changes in the phenomenal world.

Secondary qualities by virtue of being qualities as such, have no true existence, do not really reside within the object, but belong to the imagination alone. Possessing Galilean characteristics, only primary quantities have real existence as objects of thought. Having those properties of size, shape, impenetrability and motion that make them amenable to measurement, they are the hard and fast, durable stuff of creation. In Greek form, Atomism remained true to type and as speculative as ever, lacking the empirical content of modern science. Atoms only imitated their qualitative counterparts, the taste of salt for example, being the result of large jagged atoms, and fire being the result of fiery spherical atoms. This merely shifted the explanation to a different level, but nevertheless had such a hold upon Greek thought that the Atomists, along with Parmenides, were seen as responsible for the moral decline of Athens. The uncompromising materialism of Atomism encouraged moral relativism and a world of sophistry, and it was the Sophists whom Socrates took to task when he brought 'philosophy down to earth.' Or in the words of John Losee:

> Two factors weighed against any widespread acceptance of the classical version of atomism. The first factor was the uncompromising materialism of this philosophy. By explaining sensation and even thought in terms of the motions of atoms, the atomists challenged man's self-understanding. Atomism seemed to leave no place for spiritual values. Surely the values of friendship, courage, and worship cannot be reduced to the concourse of atoms. Moreover, the atomists left no place in science for considerations of purpose, whether natural or divine. [9]

•

PYTHAGORAS

Men do not know how what is at variance agrees with itself, it is the attunement of opposite tensions, like that of a bow and the lyre.

So said Heraclitus. But by vibrating the string on the bow,

plucking the lyre, you will find rational order, the measure of number and quantity, now the infamous insight of Pythagoras, who notices a remarkable thing: a one-to-one correspondence, a perfect harmony between sound and number, quantity and phenomena. It's not just approximate, it's almost exact, spot on. More than anything that could happen by pure luck or chance. It's precise, exquisite, and has a beautiful symmetry, perfect balance and measure pertaining to a new implicate order, consciousness, metamorphosis, genesis of the human mind. Because now the world 'makes sense', now there is a rational correspondence, just as mathematics 'makes sense' of sound, so Galileo 'makes sense' of the motion of a stone: for the phenomena are now seen through a new light, an 'object' truth bestowed by a shadow correspondence: a shadow not yet made a 'cause', not yet existing 'out there' as such, not yet different in kind, apart from the mind – but rather an emanation from within, an essence bespeaking the harmony of creation.

First came the Word, the Logos, now comes Number, the Logic, with a new sort of language to describe nature, as later attested by Galileo. For it is this Ratio of Pythagorean reason, the Ratio of Nature, that distinguishes the pagan from the modern, is the starting point, new twist up the mortal coil of human evolution, thus to unleash the power of science. As such, with his insights and experiment upon a lyre, it is Pythagoras who deserves to be called the true founder of the modern cave, now the mathematical matrix. For his dictum, 'all things are numbers' is going to be the defining motif, signature of science. According to Pythagoras, the author of creation is something of a divine mathematician who sings forth the birthing and becoming of the cosmos through the symmetries and harmonies of phenomena, as discerned by adepts privy to those occult signs, sacred numbers and secret algebra that are going to form the prism of modern understanding, the abracadabra of the cave, miracle of modern science, revealing horizons of undreamt control and power.

For once you have the measure of the lyre you have the measure of nature. You have the means to investigate phenomena closely and minutely in a language that gives an accurate and definite, true 'one-to-one' correspondence to reality: an abstraction of such power and precision so as to call the tune and modify phenomena — as cause to effect different modes of behaviour, different resonances, different modifications of reality in the spectrum of nature. But having abstracted this measure, is it a 'cause' in itself, this shadow correspondence: just like all the notions — the quarks, atoms and genes and molecules — those abstractions of modern science? Are these measures of causality, these shadow correspondences, the substance of reality 'out there' in any sense? Is not what we see 'behind' reality in the shadow really a co-effect, irreducibly linked to our measuring procedure? For a shadow is not a thing, not a substance, but an incidental idea: incidental to what we play upon the lyre, incidental as in a mirror whose reflections are not 'already there': for an instrument cannot play of its own just as mirror is not conscious of its reflection; for the mirror is a co-creation of consciousness: a consciousness now of infinite creative potential, now its knows the secret of measure. But taken for truth such consciousness is now the source of the fall of modern man, who falls in love with the construct, is taken by the spell, the vanity of the dark mirror. All this is to come with the deification and 'christianization' of science, but for the moment we may note that with Pythagoras, the key had been found to open the door of the cave and unlock the secrets of the universe, as Bertrand Russell is at pains to point out:

> The combination of mathematics and theology, which began with Pythagoras, characterised religious philosophy in Greece, in the Middle Ages, and in modern times down to Kant. Orphism before Pythagoras was analogous to Asiatic mystery religions. But in Plato, St Augustine, Thomas Aquinas, Descartes, Spinoza, and

Leibniz, there is an intimate blending of religion and reasoning, of moral aspiration with logical admiration of what is timeless, which comes from Pythagoras and distinguishes the intellectualized theology of Europe from the more straightforward mysticism of Asia. It is only in recent times that it has been possible to say where Pythagoras was wrong. I do not know of any other man who has been influential as he was in the sphere of thought. I say this because what appears as Platonism is, when analysed, found to be the essence of Pythagoreanism. The whole conception of an eternal world, revealed to the intellect but not to the senses, is derived from him. But for him, Christians would not have thought of Christ as the Word, but for him, theologians would not have sought logical proofs of God and immortality. But in him all this is implicit. [10]

Birthed in a cosmos bespeaking the beautiful symmetry and magical harmony of all creation, so the Ratio of mathematics was key to the Soul, the perfect language to express the unity and harmony of all things that partake in the One, which gives way to the Two, the diad, mirrored in the multiplicity of things, divided yet whole: because only through the virtue of the One do they exist at all. And each number has its own essence, correlates to a particular sound and harmony, the gateway to the Soul. Such was the mathematical mysticism of the Pythagorean school, the Mathematikoi, who had special initiation rites and observed a strict secrecy. Numbers had mystical significance and could be used for spiritual purification, thus raising the soul to union with the divine: for at the deepest level, reality is a music, a mathematical harmony. Look and you will see, the vibrations, stable patterns, from the wings of a butterfly to the leaves on a tree, slice of cucumber to the vanes of a bee, see the Sacred Symmetry, the secret Ratio, from the Milky Way to the Golden Section and Michelangelo. Don't be distracted by the object and particulars, look to the universal, the formative principle, the divine Logos, resonating all in one and one in all, as the world is literally sung into existence in the Pythagorean epiphany.

•

PLATO

From water and air of Thales and Anaximandes to the Nous of Xenophanes, from the One of Parmenides to the flux of Heraclitus and the Ratio of Pythagoras, comes the grand synthesis: Plato and his Theory of Ideas. In the Timaeus Plato described the creation of the universe by a benevolent God who imprinted a mathematical pattern upon a formless primordial matter. The Timaeus, albeit inferior to other works, was the only Platonic dialogue known to the Middle Ages. Naturally appropriated by Christian thinkers who identified with an imperfect world made of in the image of a perfect heaven, little did they know that such Platonic intrigue was arithmetic to a different sort of world and coming equation, as figured out through the Scientific Revolution.

For, according to the Greeks, the world did not come forth from the Logos or the Word, was not made in any shape or form like a literary poem or composition, that curious synchronicity of pen to paper, from thought to imagination, as the world of the Author magically appears and unfolds His Creation. Rather God, called forth creation from something rather abstract, more exact and precise, something born in the Ratio of reason, the measure of number and magic of algebra: something like 2+2=4, or the logic of a triangle, or more complex still, e equals mc squared, as the cosmos is ushered forth through a consciousness created and made in the image and likeness of another sort of God, a Greek God, something like Plato's Father, now the creator, the 'Divine Engineer' of the Scientific revolution to be. Let's read Plato's genesis again as Greek gospel to the beginnings of creation, through the Timaeus:

When the father and creator saw the creature which he had made moving and living, the created image of eternal gods, he rejoiced, and in his joy determined to make the copy still more like the original: and as this was eternal, he sought to make the universe eternal, so far as it might be. Now the nature of the ideal being was everlasting, but to bestow this attribute in its fullness upon a creature was impossible. Wherefore he resolved to have a moving image of eternity, and when he set in order the heaven, he made his image eternal but moving, according to number, while eternity itself rests in unity; and this image we call Time.

Before this, there were no days or nights. Of the eternal essence we must not say that it was or will be; only is is correct. It is implied that of the 'moving image of eternity' it is correct to say that it was and will be.

Time and the heavens came into existence at the same instant. God made the sun so that animals could learn arithmetic – without the succession of days or nights, one supposes, we should not have thought of numbers. The sight of day and night, months and years, has created knowledge of number and given us the conception of time, hence came philosophy. This is the greatest boon we owe to sight. [11]

In the beginning was the Word, the Spirit, the formative principle, breathe of intelligence. And the Word was with God. So says John in the New Testament. For Christ was the Logos: was the Word incarnate. For Plato the divine source, father and creator, is the same in action and principle but different in kind. He is logical, and over time is going to rationalise Genesis. The Timaeus mirrors Genesis, is the Logic of the Logos, thus to give a rational superstructure for a nascent church: a Logic that ends up as the travesty of faith. For having built a mathematical heaven, made according to Platonic logic and the Ratio of reason, thus to give the Christian 'Deus ex Machina' centuries later, no room was left for faith and it was kicked out the premises. For the Shadow really had no room for any qualities at all, broke the conscious connection of nature and soul. All this is implicit, and if we were to read the Timaeus as a parody of Genesis we would not go far wrong: for it was to become the stuff of serious revolution and upheaval of the world of spirit and all religion, forming the perennial conflict between the old and new in the phases of the eclipse.

Born in the abstraction of Platonic thought was the intelligence of the One, a moving geometry whose perfection was the embodiment of a circle, whose immortal symmetry was born before all worlds. At the nadir of creation is the Soul of man, whose origin is divine but has fallen to the murky depths of matter and darkness of the cave. Just as in the Bible, he is made in the image of God, and drawn in likeness to his father, the creator, made as in a mirror whose frame he cannot see: for he is blind to himself and his Soul. This world is but reflection of a higher transcendent reality that has been obscured and separated by the eye of the body which mistakes appearance for reality, an eye now ripped out and tossed into the ocean of sense, an eye that rolls around aimlessly in the world of inchoate thought and pleasure, an eye that has lost its acquaintance of nature. For the eye of our conscience is the mirror of our Soul, and we must keep it clean so the light may shine through, so says Plato:

The soul is like an eye, when resting upon that on which truth and being shine, the soul perceives and understands, and is ra-

diant with intelligence; but when turned towards the twilight world of becoming and perishing, then she has opinion only, and goes blinking about. [12]

The Soul is the eye of our intelligence, that pearl of wisdom polished bright by philosophy. This world is but a poor reflection of a far higher transcendent reality dirtied and dimmed by the senses. Our conscience is the mirror of our soul, and only be cleaning it of passion and error can the pure light of reason shine through to show the beauty of a perfect world of which ours is but a dim reflection. In the dual world of Plato there is the Universal in which the Particular partakes. In the Platonic heaven there is an ideal 'Cat' of which all other cats are but copies of the original, just as there is an ideal 'Rose', along with an infinite host of Universal beings, essential in their 'suchness' and likeness to the One, whose Beauty underlies and informs all creation, a beauty we may catch now and then in the timeless moment of a majestic sunset, a beauty that is the essence of Plato.

This Beauty is Timeless and Transcendent, truly whole and 'holy', rather than bounded, fragmented and broken. For we may think we see a 'rose' but we don't, for we have not as yet endured a Platonic relationship nor understood what it means. Because we see the rose as separate and apart, as if broken from its branch, we have not caught its essence from a realm of perfection of which the rose is but a copy of the real thing: and this copy partakes in a form that is One and Divine: the transcendent Universal Beauty, the Beauty of a Rose we may recognise in our better moments, a beauty lost but now remembered, there forever, should we clean our eyes and see life as it is: now transcendent, divine.

All this derives from something simpler, Plato's obsession with the most perfect form, the circle. For just as there is no perfect rose in nature, so there is no perfect circle: such an ideal form does not exist. But without this good Circle, this perfect conception, we would never be able to know any circle, for to know something is to literally 're-cognise' it, see a likeness to something prior, already there. We cannot know a rose or a circle or anything all without knowing them in some sense, through some semblance before, and what we see, what is presented to the senses, comes after, as a blurred imperfect copy: blurred and imperfect because we live in a mixed up reality, a seeming chaos. What we see comes through our intelligence. What we see is that which is 'intelligible' to us. What we see is made real; is 'realised' through that intelligence – an intelligence which comes prior, comes first, comes from within, now the Soul: the Divine Intelligence informing all, and this Intelligence has fashioned our minds as like itself, in the very same image, so we may recognise the Divine Transcendence of creation, as 're-cognised' in the immortal symmetry of a radiant sunset, that timeless moment when the veil is lifted and all is One, simply that which 'IS', and we get an insight into the Good, the Just and the Beautiful, the Soul of creation – and the mind of the Greek.

As in Eastern philosophy the manifest separateness of the world is an illusion, but with a peculiar Greek twist, math-mystical conclusion: for the Divine Intelligence is the Ratio, the One in All, is the universal formative principle, eternal fountain, wellspring of creation. Now the closest fit of particulars to universals is expressed by the timeless purity of mathematics, whereby everything in nature takes the shortest path and forms a perfect isomorphism to everything else: shows a 'one-to-one' correspondence that is abstract, essentially the same, a place where the introduction of sense perception cannot take part: for to do so would give it a quality: a blurred imperfect copy. This quantitative realm is bears little relation to reality as we perceive it: rather it is prior, higher and above, a hidden unseen relation behind the apparent chaos. In this view the mathematical formalism of the angles of a triangle, or 2+2=4, is genuine knowledge because it comes prior, and must do so by necessity:

for there must be some stable pattern in nature, some frame upon which we may hang existence. As such only mathematics can really express the formal truths of universal relations, and these relations, because of their timeless purity and abstract perfection, are the background against which the world is but a stage, a play. Its truths can be nothing other than what they are. No other truth has such beatific certainty.

And such certainty Plato establishes through a dualism that is become the foundation of the modern Matrix to come, as he assigns primary to secondary qualities, the primary realm now the moving geometry – modern mechanism to be – behind the original Greek elements. Parmenides and Atomism and Pythagoras all concur, the mathematics of pure geometry replacing atoms in motion as the underlying substance of the universe. The old Greek elements are correlated to geometry, for now nature is subordinate to a pure geometry, a solid state symphony, the keystone of Platonic thought. Pythagoras is now made all bi-sectional, angular, tangential and obtuse in the Platonic theory of solids. Earth is assigned the cube, the most solid of structural geometry, and the Atomic fire is now given a mathematical form: no longer vaguely spherical, but now precise, exact, taking on the form of a tetrahedron because of its sharp, fiery angles.

As yet the calculus is not there, there no truth correspondence, no Newton, and Greek philosophy remains true to type, as speculative as ever, and divinely, blissfully so. Archimedes is yet to come, there is no experiment, Plato would not truck with matter and preferred to remain in a geometric heaven unsullied by the senses, nor did it occur to him to test his principles in the 'real world', drop balls down inclined planes as did Galileo, as a gentle introduction to the New Science. Essentially his vision was sublime: being is but the becoming of perfection in remembrance of the Divine. And this divinity is not so much a loving being but a beautiful intelligence, not so much a presence but prescience: not so much of the heart but of the head, an intellective act, cerebral – a Prescience now the Soul, the formative principle, the divine eye in us all, whose immortal symmetry is the wherewithal, the true basis for perception in everything we sense and see. For you are what you think. And as you learn to think in accord with the Divine Intelligence so everything is made beautiful. Look within, says Plato, look inside the eye of your Soul and recognise its active principle, reflect upon it and you will come to know yourself, the wonder of nature, and all true philosophy.

But you must learnt the language of true forms. It's geometry, mathematics: the true correspondence to reality, mirror of nature: the invisible hand of God, the formative principle behind everything – repeatedly endlessly and contiguously throughout matter. 'Geometry', says Plato, 'will draw the soul towards truth, and create the spirit of philosophy.' Dealing with pure abstractions – points, lines, angles – geometry is the body politic unsullied by sensuality, undaunted by opinion. For philosopher kings shall rule Plato's city of god, and they shall not be subject to the ad hoc or democracy of the Athenian mob, nor the Sophist or demagogue: for they shall form the Republic, founded upon rock solid logic, and these kings shall be trained for ten years in mathematical arts such as arithmetic, astronomy and harmonics. For arithmetic says Plato, 'has a very great and elevating effect, compelling the soul to reason about abstract numbers, and rebelling against the introduction of visible and tangible objects in to the argument.' For like Parmenides, Plato is Greek through and through, you reason yourself to the truth. For by beginning with self evident axioms, you may construct the body politic, even your very own cosmos, starting at the apex of the triangle as everything corresponds and follows suit.

Such an approach was about to play pandemonium with the medieval cosmos, create the great modern world systems, great cathedrals and castles in the air. From the British Empire, all liberal and empirical, from Aquinas to Kant, from the spiritual heights of the cathedral of Chartres to the rarefied air of the gas chamber, from the French guillotine to the New Atlantis of America, from Hegel to Marx, Darwin to Freud, Stalin or Hitler, you could do anything with a pencil and rubber and pair of dividers – without the introduction of visible and tangible objects, the rub of humanity.

•

ARISTOTLE

If Socrates had planted the seed, and brought Greek philosophy down to earth, then the final germination came from Plato's long-standing student and heir: the first formal scientist, cultivated naturalist, grounded in observation. If Plato lived in a world of mystical idealism, then his pupil was very particular about exactly what sort of world this was, and how it was derived from experience. So Aristotle filled in all the particulars, all the details and taxonomy, rooting out all the weeds, all the accidental correlations so as to establish proper categories of truth through observations and explanatory principles in that inductive-deductive procedure which was to become the modus operandi of modern science. The cave was put in order, all neat and precise, modelled upon the Greek infatuation with geometry. Starting with the lucidity of self evident axioms, geometry reflected the beautiful consistency of nature, and perfection of the Divine Mind. And so from simplex archetypes of Logic came the ordered landscape of the Cave, as derived by the methodical patience of Aristotle, who reaped the fruits left by Plato.

For Plato, the fact that there are no perfect circles, squares or triangles in the natural world is irrelevant, entirely incidental – for the world as such, he would argue, is incidental. Just as all the rays of light that enter a mirror are incidental and deal only with the exteriority of things as seen from the 'outside', so the mirror of consciousness is but a soapy bubble of dreamy and confusing forms. If you really want to make true contact with reality, then take a look in the eye of your mind, the platonic lens, and your will see the timeless purity of abstract relations. For Aristotle these first Causes, these primary relations, cannot come from nothing, but are derived from experience, from nature. And nature's midwife is man who induces and draws out sensible truths that bear witness to the birth and becoming of the cosmos. The ideals of geometry may in some sense be 'given' but they are nevertheless induced, drawn from sense experience, as observed by the pedestrian patience of Aristotle. Only by drawing upon the phenomenal world can we glean those necessary logical relations underlying the natural order, for the principles underlying the cosmos are rooted in rational observation and sense experience. Reason is rarefied through the spectrum of nature and truth is made crystal clear through its axioms. Just a refraction through a prism is a two way process so is the study of nature a dynamic interplay of the inductive-deductive procedure.

And so the Universals of Plato become the Forms of Aristotle as induced from sense experience: only by beginning at the bottom can we get to the top, each little step not only categorical, but ascending, and properly corresponding, to nature so as to form a pyramid of knowledge, infrastructure for science. Our cave now has an implicate order, an operational method and precise definitions, for only by looking at particulars can we draw conclusions, only be inducing the facts of experience can we glean those self evident axioms necessary for truth to exist at all.

Aristotle crystallized Plato's metaphysics into a working appliance, an inductive-deductive approach, now founded upon the regulative principles of reason, a body of systematic knowledge later to be called Newton's Method of Analysis and Synthesis. As such Aristotle deserves to be called the Socratic midwife of modern science. He is Plato applied, analytical, methodical, pedestrian, often-times pedantic – but magnificent. With him the aperture of the cave opens up to a whole new spectrum of nature, as subjects such as motion, optics, and astronomy acquire axioms and definitions, and biology gets a whole new taxonomy, as Aristotle, prelude to Darwin and working in his favourite field, and much like the Victorians to come, manages to categorize, assimilate and pin down hundreds of species.

For just as there is a necessary correlation of the angles of a triangle, so there is in nature. Science begins with the understanding that certain events and properties correlate and co-exist, and its object is to isolate them and refract them from the chaotic spectrum of experience. The facts of experience, just like the lines of a triangle, are necessarily related to each other. We may induce that co-joining of three lines of a triangle make up a figure, and classify it a triangle, but as yet we do not know its logical relations. What we need is a formula, an axiom system, made up of definitions and theorems from which we can deduce these relations. Geometry forms the purest science because there is a one-to-one correspondence of its axioms and laws to reality. For each and every logical process there is a corresponding link to our measuring procedure.

Geometrical laws may be fixed and immutable, but life itself is evidently not. For consciousness always changes form, and what comes to mind and appears to thought, what is given to the senses as sound and colour, touch and texture – is not exactly likened to a triangle. All this though is an illusion, for what lies behind phenomena is a pattern, a hidden unseen relation, an implicate order, co-related like triangle. Cooly observe, induce the facts, derive explanatory principles, join the dots, and in the analysis and synthesis you may observe the motion of the stars, roll balls down an inclined plane, put gas under pressure, weights on a lever, vary the conditions, watch the correlations, and thereby derive consistent equations, testable and repeatable under experimental conditions.

Such is the light of science as rarefied from sense experience, such is the prism of our understanding, all linear and logical. For if we care to take a look, there is a dazzling crystal up and behind us in the far recesses of the cave, and it comes from that Golden Dawn bequeathed by the Greeks, a dawn that now turns to dusk, a dawn that sleeps for a thousand years before re-awakening in that spring and flowering of knowledge that was Renaissance. As Greeks works, once lost, are retrieved from the East, and translations make their way back to European shores, so the light returns, and spring turns to summer, not just a re-birth now, but the final reaping of the fruits of the harvest, as the Greeks descend to rule the world and finish the job where they left off – in that wonder called the Scientific Revolution.

• • •

Revolution in Science

'I perceive,' said the Countess,

'philosophy is now become very Mechanical.'

'So Mechanical,' said I, 'that I fear we shall

quickly be ashsam'd of it: they have the World

to be in great, what a watch is in little;

which is very regular, & depends only

upon the just disposing of the

several parts of the movement.

But pray tell me, Madam, had you not formerly

a more sublime of idea or the Universe?'

Fontenelle

The Plurality of Worlds

REVOLUTION IN SCIENCE

THE DARK AGES • AQUINAS
COPERNICUS • KEPLER • GALILEO
DESCARTES • NEWTON

•

THE DARK AGES

Before the Renaissance and the Scientific Revolution comes the Dark Ages of the West. Poised in Hellenic logic, denying the visible world as false and illusive, Greek thought provided a metaphysical springboard that once drawn back would unleash the power of science. What is surprising about the Greeks is not just the depth, but the sheer variety of their thought. From Heraclitus to Parmenides, to Pythagoras, Plato and Aristotle, the Stoics, Cynics and Sceptics alike, all may have been contrary, each different, but all were inspired by one passion: the Socratic spirit and its free quest for truth. The dissolution of this Golden Age was reflected in the political entropy of the Roman Empire, the centre could not hold and so gave way. There was no one faith or power to unify the whole, to take hold of the world in its entirety and throw it into a single truth, the substance of world history. This was the task of Christian theology.

If we look at the preamble to the Mechanical Philosophy during the Renaissance, its encyclopaedic knowledge reads as a natural continuation of the exuberant curiosity of the Greeks. What is strange it that it took over thousand years before Greek thought, in all its glory, would be revived. From Aristotle onwards Greek philosophy fell into slow decline, and we can draw a symbolic line between its ancient death and modern rebirth. With the closure of Plato's Academy in 527 AD the doors of philosophy remained firmly shut, leaving a few burning embers, chinks of light, until they were finally opened again in 1543 by Copernicus, who trumpets the new age with Plato's immortal words:

Let no one ignorant of geometry enter here.

For the fair phrase once written above the entrance of Plato's Academy is now resurrected on the frontispiece of *On the Revolution of the Heavenly Spheres*, a little book written by a timid Copernicus, who had the temerity to announce a new age and cosmological order – albeit on his deathbed.

After the Golden Age of the Greeks and the power and might of the Roman empire, the world had gone back to default, it was business as usual, chaos and political upheaval – all from which the Catholic Church would remain triumphant. The closure of Plato's Academy announced the sleep of reason, and as the Logic of the Greeks was eclipsed by the Logos of the Bible in the dark cavern of Christendom, so the Gospels became the final simplex Word in all matters, manners and discourse.

Rome became Christian and held the key to salvation: excommunicated emperors, brought kings to their knees, blessed the pious, whipped up the rent, charged for confession, thus to exorcise the world of sin through the blood of crusades and torture of inquisition. The hope of the age was now pinned upon the cross, and heaped below in fire of the cavern were the busts, bellies and not least brains of Greek civilization. For Rome was now the City of God and all roads led to the Cross, as expressed by the greatest of all Catholic theologians, Saint Augustine:

When, then, the question is asked what we are to believe in regard to religion, it is not necessary to probe into the nature of things, as was done by those whom the Greeks call physici; nor need we be in alarm lest the Christian should be ignorant of the force and number of the elements; the motion, and order, and the eclipses of the heavenly bodies; the form of the

*heavens, the species and natures of the animals, plants, stones,
fountains, rivers and mountains; about chronology and dis-
tances; the signs of coming storms; and a thousand other things
which those philosophers either have found out, or think they
have found out … It is enough for the Christian to believe that
the only cause of created things, whether heavenly or earthly,
whether invisible or invisible, is the goodness of the Creator, the
one true God; and that nothing exists but Himself that does not
derive its existence from Him.* [1]

The City of God was the Bible of the Dark Ages, and enlightenment lay beyond the realm of Rome in the remote con-
fines of the old Eastern empire now decamped in Constantino-
ple, where the remains of Greek works were kept in safe storage
in its famous library, away from vandals. Other remote boundaries remained beyond censure, such as Ireland, where in the
ninth century the remarkable John the Scott could set reason
above faith, dismiss ecclesiastical authority and enjoy his own
little Renaissance.

In the meantime Rome set about creating those dry ricks
of ignorance and distinctions of body and soul which would
inflame faith and set the European spirit alight once sparked
by reason. The Greeks had philosophy but no god. Christianity
had god but no philosophy, and in retrospect, it was only a matter of time before Genesis and Timaeus walked hand in hand.
One can see the attraction, as long, of course, as the Logos had
the upper hand and last say, the Alpha and Omega, first and last
word – on everything. For the book of nature was the Bible, and
its language was written in a native syntax: a symbolism pertaining to qualitative relations and what we know through the senses – in short plain English, or rather: Latin. And the Church's
word was the Word, the Logos, and because it came from God,
it had the last word in the matter. The Timaeus though was
a rudimentary guide, instruction manual, a pocket logic that
explained a lot, filled in all the gaps, as to how exactly it all

worked. Trouble was, it was written in a different language with a different set of values that got lost in translation: was written in mathematics, a mathematics that described a world different to the senses, different in kind, abstract and quantitative, all algebra and logic – or in plain Platonic terms: Number. In time each would be fighting its own corner, neither being able to understand the other, nor see the absurdity of the argument: for they did not even speak the same language.

For she the Church was in want of a husband around the house to set up shop, and very useful he was too, very obliging, not only as a practical aid but very reasonable too, for a mighty fortress indeed was to be the City of God: founded upon self evident principles and rational constructs. Very useful indeed was the Greek, only as long as he was kept upon a ball and chain, anchored to the rock of Rome, and did not wander off into the hills and vales of nature and start coming up with his own conclusions to upset the Holy and Apostolic. For the Greek was a dogged philosopher, and dogs, being what they are, will do what dogs do best, and through an instinctive logic, will sniff around the primordial tree and draw their own conclusions so as to outperform their duty and demonstrate their brilliance – as would Galileo – who was about to drop the Church right in it.

For the philosophical pups of the revolution conceived by the marriage of Greek and Christian ideas were different in kind – actually born deaf, dumb and blind, were quite something else, so much so as to be unrecognisable by their parents. For from the *Logic of the Logos* came a far deeper distinction between spirit and matter than had ever been conceived by the Bible or the Greeks. The Biblical picture of an almighty God acting by law upon nature from the 'outside', once rarefied by the Greek prism, implied a rigid determinism that was going to turn the Medieval world inside out. As Christendom absorbed the message as freshly delivered from Arabian shores through translations of the Greeks and Aristotle, the Logos became Logic in exact proportion to the newly emerging faith in a mechanical creationism. As refined by the ready-made ideals of the New Philosophy, so faith found it had bitten off more than it could chew, and before long it was found wanting, divested of meaning, choking on truth.

The Logos was the final Word on all knowledge and discourse, the beginning and end, Alpha and Omega, of all human aspirations. Its empire was founded on the rock of Rome. That this rock still remains embedded within the world's conscience is tantamount to its remarkable fortitude through the vicissitudes of time. The Catholic Church has a distinct caste of priesthood and theology, remaining essentially unchanged since its inception, standing out as a beacon to light up the landmarks of a distant past. Meanwhile science has moved on.

Catholicism has stuck to the Bible, Mechanism to mathematics. Each claims an objective description of the universe from qualitatively different ways of seeing: one from the Logos of the Bible, the other from the Logic of Science, each abiding in two separate worlds, one circled by faith, the other squared by reason. One stands still, the other advances, one looks to the past, the other to the future, one is behind, the other ahead. For rather like Zeno's tortoise, the slow contemplative, timeless and patient wisdom of religion out stands the restless run of science. For, after all its ceaseless striving, all its deep insights, mighty force and energy, all its talk about matter, all the analysis, sophistry and patter – all is in vain, for at the end of the race and chequered past lies what religion says was there in the first place: the way and the truth of Christ, the real end and meaning and purpose of life.

Without Spirit, religion would claim, life is just what science proposes: the mere hurrying of a mechanical hare chased by the dogs of philosophy, the irony being that the hare and the whole hunt and chase for truth was set in motion by a Biblical god, for once Christianity went Greek it would always find reason snapping at its heels, the Logic now chasing the Logos, just itching to find out how God made the world. For in the beginning, when God said 'Let there be', what was the logical counterpart? As reason chased faith so it got caught up and ensnared in its own axioms: for the hare was after all nothing but a mechanism set there by God since the beginning of creation. Reason could never quite get ahead because faith kept it going and led it on: for the answer was 'out there' somewhere. And try as it may to find the final solution, the answer would always elude this Greek dog of philosophy as it found itself chasing a lost cause and lapping the crossed-sort logic of its mind.

This mechanical animal was the most peculiar contribution of Christian theology to the modern world, for if we conceal our axioms of faith within a material agency other than our own we are to be forever chained by its logical necessity. For into The City of God was brought a Trojan horse: something that would betray the prince and kingdom, leaving Beauty to sleep for centuries to come. Hidden and concealed within the premises was a fatal pinprick that would pop the Medieval bubble and prove the downfall of Christianity and all spirituality, leaving the souls of Christendom forever chained in another sort of cave and new sort of religion.

•

AQUINAS

The pinprick was innocuous enough. It was administered by Aquinas, who would never live to see the results, taste his own medicine. It seemed like a good idea at the time, had arrived on European shores from Arabic sources at the turn of the millennium, all in a ready-made bottle called Aristotle. It was heady stuff, a dose of salts that was to excite the Renaissance, the Greeks having now been rediscovered from the East, which still flourished, as we have noted, in classical learning. The medicine was duly proven by Aquinas, the Angelic Doctor, who held it up in the light of Biblical revelation and christened it: for the philosophy of Aristotle led not to the god of Aristotle but the God of the Bible, who was about to go all geometrical. For the Medieval window upon faith was about to be refracted by a linear light of admirable precision into a new catechism of existence. Coolly administered in the pinprick was a drug that was to cause an incurable habit, actually an addiction: for from the assumption that a metaphysical principle is a god and god is a metaphysical principle was derived a new sort of religion, drawn in kaleidoscopic certainty, a moving geometry, not yet without colour, but soon to be made uniform and pale into qualitative insignificance.

Aquinas set out to prove that God exists in exactly the same way one would prove the logical relations of a triangle, and what he provides is the cause, the Prime Mover, now actualised by a Biblical God, a now separate, determinate creator who works upon creation from the 'outside'. For the 'One' now not only has a definite existence, (in the terms of Aquinas 'He who is'), but a will which sustains and actualises reality in every act and instant. He is the First Cause of what is to become a geometry that moves. And what is a geometry that moves? Nothing more than a machine, whose every movement in linked in a one-to-one correspondence to reality, nothing more than a divine automaton now regulated by a Logic which now operates on every element of the finite process, initiates motion and sustains, being the first, efficient and final cause of creation as defined by Aquinas' famous 'Five Ways' of demonstration.

Hauling the Aristotle on board and blending him with Christian thought was no mean feat, lest we forget what was inside Aristotle's bottle was heady stuff, a Socratic spirit quite unkeeping with any pagan preconceptions or the closed precepts of the Church. Blending Aristotle with the Bible into the *Summa Theologica* was not really going to work, and the Greek genie to come out the bottle was no Saint, more a creative genius, a Galileo or Michelangelo.

For if we assume nature is triangle, a triangle we shall find, and as pure logical relations cannot be contradicted, conveniently ostracising any form of dissent, the territory is won, made over to a new likeness. Unfortunately for Aquinas non of this was apparent, and his pious hope that reason was an aid to faith only ended up with it becoming the apostasy of religion – not just for Christianity, but any form of spirituality. For, after all, by his own approach and definition, the way, the truth and life had become rather abstract, and the entry to heaven was indeed straight and narrow: a straight line, all geometrical, all very Newton and Galileo, who are waiting at the sides, ready to make their entrance so as to draw the curtains.

The Church sniffed the Aristotelian brew, and after initial suspicion and circuitous hostility, wagged its tale, and called the entire herd to pasture. Unable to see the ruse nor raise heads above the horizon, so the herd was quite oblivious to the spiritual precipice, the abyss to come. Europe had awoken from the Dark Ages in a new frame of mind, brand new cosmos, courtesy of the Greeks. Rome had bought the picture, coughed up at a price it would never be able to contemplate, let alone pay, and to which the world would be held to ransom till the end of its days. Appended to Thomist metaphysics came a Christian adaption of Aristotle's play of the universe. This Medieval cosmological bubble, held together by the 'Great Chain of Being', did not last long, and was soon to burst under the very existential pressures it had created. All it needed was a pinprick to pop the illusion, expose the fallacy. The Church, having underpinned its faith in Aristotle, could only watch its authority disappear as first the Renaissance, then the Reformation, and finally the Scientific Revolution, finished off the very process it had set on motion: the complete metamorphosis of the Logos in the light of a mechanical creationism.

Thomism itself formed a closed and static union of Aristotle and the Church, but it inherited an itching curiosity that was to irk the status quo. Born of natural questioning, Aristotelian method recreated a spirit of independence free to draw conclusions without religious dogma: and the problems of mixing Greek method and Christian dogma did not go unnoticed. For Franciscans such as Duns Scotus and William of Ockham, studying in the new universities of Paris and Oxford, the combination of geometry and faith was a little too facile. For Christian faith to remain pure it must remain true to itself and not the appendage of a plagiarised Greek god tacked onto the Bible.

Reason was independent of revelation but as it moved away it had to break through the old corrupt fabric of the Church. Wycliffe called the Pope the Antichrist, Luther nailed his complaints, both incensed by all the hypocrisy and sanctimonious luxury. Calvin called for a return to purity, Dante wrote parody, and Leonardo became the new sort of hero: not saint, but genius – and by the end of the fifteenth century this generation had its own press to follow Columbus across to the World. The little world of Medieval faith was rapidly disappearing to reveal the vast horizons of the New World, together with a new found confidence to chart and conquer new territory. Catholic reactionism merely incensed pride and fanned the flames, and even as late as 1600 Geordino Bruno was burnt at the stake for being a pernicious pain and very unAristotle, actually infidel: for had the unfortunate intelligence and temerity to foresee, bear witness to, and openly declare what was to come: an infinite universe, and the plurality of worlds of a modified Atomism.

Yet the bubbly marriage of the Greek and Church remained intact, and looking at Aristotle, one can see the beauty and attraction, why Rome bought the picture. A magnificent system of organic wonder, the Aristotelian universe was divided by the moon into two zones, the celestial and the terrestrial. Around a stationary earth revolved the stars, attached to the inner surface of a rotating sphere. The celestial zone, all the stars, moved around in perfect circles reflecting the timeless perfection of the heavens. This Platonic symphony was built, layer by layer, upon concentric spheres, encompassing first the planets and stars, and just above was the special club, the heavenly abode of trumpet angels, all the seraphims and cherubims – all motivated by the 'Prime Mover' of the great cosmic clock.

This all suited the Church nicely, was a pretty good reflection of divine perfection, especially as down here on earth, according to Aristotle, there was a lot of muck, in contrast to the timeless perfection of the heavens. Change remained qualitative, and the earth was made up of not abstract, mathematical entities, but sensible and qualitative things, the elements of earth, air, fire, and water, which all occupied their 'natural places', according to their respective properties. Earth being the heaviest element, was drawn toward the centre, surrounded by water, then air and fire. Then God, through the Prime Mover, introduced motion to the universe, first through the stars, which whirled the planets, down through the spheres to the moon and elements, which got all stirred up in the communicative process. God though intended the universe to return to its natural state, so everything tried to get back to where it properly belonged. A stone gravitated toward to earth – in a straight line, the shortest route possible – to return to its natural place, and fire surged upwards in the form of flames and volcanoes because that was likewise its tendency. All this sort of 'qualitative' thinking was to irk Galileo who suggested it explained nothing: for if you really want to know how a stone falls to earth, get out a water-clock, pencil and paper, and drop it from the Tower of Pisa.

In this Great Chain of Being, everything from the lowliest limpet to starry sky, had its proper place in the heavenly hierarchy, each form of life being actualised by its inherent potential for perfection as set by God since the origins of creation. And just as a stone was obliged to behave perfectly according to its potential form by falling to the ground, so man's proper inclination is an ascent predestined by his own unique place at the top of creation. Made in the image of the Divine, and by virtue of his Free Will and Soul, his natural resting place was in Heaven with God himself in the abode of Angels, the place where he actually belonged before the Fall. La belle fontaine, the wellspring of causality, so God filled and united all creation.

Life may have been brutish and short, but the Medieval world, in comparison to the vast impersonal spaces of the modern universe, was a cosy cosmos motivated, much like the Greeks, by poetic meaning and archetypes of Love as much as war, bigotry and violence: for it was one unified whole, one living spirit, comprehended by a personal yet transcendent creator. The earth, the terrestrial zone, was naturally associated, through nurture and reproduction, as female, as sensitive, alive, and purposeful, a womb impregnated with a living rational soul. Things were known through their aesthetic rather than physical characteristics, through their verisimilitudes and harmonies, through their qualitative 'sympathies' embodied in 'occult' sciences such like medicine and astrology. Plants had ruling planets, metals grew and regenerated like animal veins, spotted herbs cured spots, and yellow ones, jaundice.

The Medieval attitude toward death and regeneration, to spirit and nature, also had a dark side, reflecting ambivalent attitudes towards women. Nature could be a virgin, and to offend her was tantamount to rape, or a harlot from hell, ready for the flames. After all it was Eve who had been tempted by the serpent and had offered Adam the apple. Likewise the body of the Church could be corrupted by pagan nature and the seductions of the flesh in the Medieval marriage of heaven and hell. A witch was a fallen women in league with the devil, who by her seductive sorcery, lead the descent into hades. Carnal pleasures were evidence of witchcraft, and demanded little pocket detective guides such as *The Copulation of the Devil with Male and Female Witches*, escorting the perplexed to those little tell-tale marks on the labia.

Lest we forget the great century of witch burning and the Scientific Revolution went hand in hand, was occult and penetrate to what was to come, as suggested by Francis Bacon, the great propagandist of the New Science, inspiration to the new

Royal Society, as he called for the reconquest of nature after the Fall, for she shall be 'put in constraint, moulded and made', according to the artifice of man. But before the aims of the Royal Society could be realized the old cosmos had to die, the acid test coming with Galileo's infamous trial.

First, the spiritual clockwork embodied in the Greek obsession with circles would have to be rendered precise, quantitative and mechanical so as communicate a new message devolved from above unto those below, communicated down not so much through concentric spheres but an intricate machinery of cogs and wheels. A universal physics rather than soul was really more appropriate to understanding the mind of God. Clocks and mechanical contrivances were already familiar to the Medieval mind, with grindstones, waterwheels, pumps and pulleys appearing as welcome signs of a new economy and control over nature. As early as 1232 the Sultan of Damascus had been presented with a clock imitating the motions of the sun and moon, and in the fourteenth century we find Nicolas of Oresme, bishop mathematician, already comparing God to a Divine Clockmaker:

> *And these powers are so moderated, tempered, and ordered against their resistances that the movements are made without violence. And except for the lack of violence it is like the situation when a man has made a clock and lets it go and be moved by itself. Thus it was that God let the heavens be moved continually according to the proportions that the moving powers have to their resistances and according to established order.* [2]

As re-serviced by the New Astronomy the old cosmological clock would have its face rearranged and made to a new order, with inefficient notions of spirit replaced by force, and so engineered to obey perfect mathematical laws, cleaning out all the old occult cobwebs.

•

COPERNICUS

The Mechanical Revolution, like anything novel or original, was painstakingly slow and secretive, accompanied by a fear of being seen to challenge the status quo, the powers that be, God's truth and ways of the Church. It centre spring was wound up innocuously enough by Copernicus. His heliocentrism, sun-centred system, was not new, nor unprecedented or novel – that privilege, as always, belonged to the Greeks: both Empodecles and Aristarchus of Samos had suggested it, nor had they attracted much attention or charges of impiety.

But it was fear of very such charges that kept Copernicus from publishing his *Revolutions of the Heavenly Orbs* for over thirty years, the first completed copy reaching him hours before his death in 1543: and then again it attracted little interest. Characterised by an intense secrecy, Copernicus lived in fear of authority, despite evidence to the contrary – at one stage even Rome urged him to publication.

Caution though was the byword in all matters of faith and reason. To be left solely responsible for contriving a publication contrary to established reason was not exactly prudent. Copernicus knew his theory was rather uncertain, with few original observations, relying heavily on the same, self-made tools of the Greeks and their age old tabulations. He dreaded the mockery of his peers, and had no wish to combine genius with martyrdom, customarily reminded that the habit of heretics tended towards flames. Secrecy, piety and deference toward papal authority sufficed, changing as it did from liberalism to reactionism according to the whims of the Vatican. Fame, then, lay posthumously bound on his deathbed, beyond reproach, just when he could no longer read nor care. During his lifetime Copernicus had become the enigma of a generation, a subject of curiosity, hiding away in his observation tower with his manuscript. In death his legend began, immortalised through his successors: Kepler, Galileo, Descartes and Newton.

•

KEPLER

Copernicus marks the end of The Dark Ages, and after a thousand years the doors of Plato's Academy are open again, with those familiar words written on the frontispiece of the *De Revolutionibus Orbium Coelestium*, announcing the dawn of a new age:

> *Let no one ignorant of geometry here.*

Enter Kepler, who felt quite at home, was in entire accord:

> *Why waste words? Geometry existed before creation, is co-eternal with the mind of God,* **is God himself** *(what exists in God that is not God himself?); geometry provided God with a model for creation and was implanted into man, together with God's own likeness — and not merely conveyed to his mind through the eyes.* [3]

Indeed. Why waste words? For geometry now is not merely Platonic, but gone Christian: *is God himself.* For like Kepler we must have it in italics. God himself being the First Cause, now the Shadow to be cast in the object-light. Not that Kepler nor anyone living betwixt the Middle Ages, mathematician modern or otherwise, can be blamed for seeing his pet subject with divine eyes. Mercurial in mind, afflicted by a terrible childhood, encumbered by lifelong medieval maladies, Johannes Kepler, much like Copernicus, his forbearer, sought sanctuary in the Pythagorean perfection of a clockwork astronomy. Just as Plato had done with his elements, so he proudly announced a new harmony of the heavens neatly nestled in the five regular solids, achieving a rough agreement with the radii of planets. He used Copernicus's ancient tabulations, but later the figures of his grandiose *Mysterium Cosmographicum*, published in 1596, were no longer to be found squeezing into their facts. With typical reluctance Kepler eventually abandoned his project, his faith unshaken, finding final consummation through the *Astronomia Nova* in 1609. Years of passionate Pythagorean perseverance and solitary struggle yielded the first laws upon which Newton built the universe:

> *My aim is to show that the heavenly machine is not a kind of divine living being, but a kind of clockwork, (and he who believes that a clock has a soul, attributes the maker's glory to the work), insofar as nearly all the manifold motions are caused by a most simple, magnetic, and material force, just as all motions of the clock are caused by simple weight. And I also show how these physical causes are to be given numerical and geometrical expression.* [4]

And such geometrical expression yielded modern fruits, such as the beautiful economy of his most celebrated Third Law: that the ratio of the planets is proportional to the cubes of their mean distance from the sun. For angels are to become angles, souls – forces, spirits – substances. Aristotle was mechanised, the sun the centre of the clock, no longer to be given impetus by the soul, but by a force that Newton would later formulate as gravity.

> *If we substitute for the word 'soul' the word 'force', then we get just the principle which underlies my physics of the skies in the Astronomia Nova ... for once I firmly believed that the motive force of a planet was a soul ... yet as I reflected that this cause of motion diminishes in proportion to distance, just as the light of the sun diminishes in proportion from the sun, I came to the conclusion that this force must be something substantial.* [5]

•

GALILEO

In 1610, just a year after the publication of his new astronomy (and ten years before the accusation of witchcraft and torture of his mother), Kepler received some exciting news. A Paduan professor had turned a spyglass upon the heavens thus to observe the moon and discover four new planets. Galileo Galilei had arrived upon the scene and he had been playing with his new toy, a little too close for comfort for the neighbouring Vatican. Gazing from his technological masterpiece he coolly noted his observations in *The Messenger from the Stars*, 'with all the certainty of sense evidence that the moon is not robed in a smooth polished surface but is in fact rough and uneven, covered everywhere, just like the earth's surface, with huge prominences, deep valleys and chasms.'

God was fickle. If the moon, the divine pearl of the Aristotelian cosmos was mere matter, a mere meteor in motion, then why not the planets and stars? Not for the first time modern technology would contradict faith and metaphysics, and the challenge was on. Either this new evidence of the senses was to be accepted, thereby calling for a new truth, or it was to be deemed inadmissible by poor criticism, or simply a refusal to glance – as some did – through a magical contraption that threatened to bewitch centuries of truth. The moon had never been circumscribed this way, no longer sublime but material, and the Church was understandably guarded over the matter. At stake was the whole modus operandi of the Medieval cosmos: break one link in the Great Chain of Being, and the whole thing would fall apart. And the Vatican would soon find itself hemmed in, circumscribed by a moon which was now not just a rock but a wobbly one at that, soon to muck around in an egg shaped orbit, its elements thrown all over the heavens. Sympathies were strained, links were broken, no longer fitted facts, the universe lost its chain of reasoning, all coherence gone.

Galileo obliged. Now what united the moon to the lowliest stone and highest star was not a universal soul but universal physics, or more accurately, the force of balls in motion: and these were about to be aimed into the window of faith with impeccable logic and precision: for unlike the Church, which was in the habit of choosing those philosophical fruits that suited its needs and happened along the way, Galileo had really done his Greek homework, and had come to conclusions not so palatable to established wisdom. And the fact that the window of faith was a magnificent construct, beautiful work of art, had taken generations to construct, mattered not one jot: for it was about to be not just shattered, but annihilated by the New Science.

For just desserts had come from mixing a Biblical god with the Greeks, desserts now divested of essence, divinely indigestible, leaving an unsavoury taste, to which Descartes would add a dash of salt and Newton the poison apple. For a new philosophical fruit had fallen from the tree to give a new sort of Adam and Eve, born not so much of Genesis but the Timaeus, and as they enter the new abode so Galileo opens the door and is rubbing his hands, juggling recipes, ready to serve customers, cooking up ideas and running the errand of experiment, all infantry to an uncompromising materialism.

What mattered to him was not so much that the moon was terrestrial, but rather the reality behind it, the substance underlying its qualitative and sensory appearance. And this reality is what made him suspect, a heretic to the Church: for it existed in a territory so alien to the Medieval mind it might as well have come from another planet. For these new creatures, these New Philosophers, were like aliens from afar, were *different in kind* to anything that had come before, and they were about to invade the earth and rewrite its history, or rather the book of nature, a nature now not seen through the constructs of spirit and consciousness, heart or soul, but something rather different, utterly abstract, and the Church Catholic was caught in the middle to muddle through the act. For it had the misfortune as to happen to be there, bear witness to and mitigate, aid and abet, the greatest revolution ever to grace the human race: for whether by chance or design, destiny or fate, came a new countenance divine, genesis of mankind, born of the Christian faith. Not that the Church really had a clue as to how to even begin to approach the New Philosophy, for it was entirely new and foreign territory, and it was having trouble understanding the language, let alone dialect. For the book of nature was not written by the Logos of the Bible, but the Logic of the Timaeus, now mirror to Genesis, as seen through mathematical relationships, with Galileo the new Saint Peter taking over the keys to creation – the key being mathematics:

> *Philosophy is written in this grand book – I mean the universe – which stands continually open to our gaze, but it cannot be understood unless one first learns to comprehend the language and interpret the characters on which it is written. It is written in the language of mathematics, and its characters are triangles, circles, and other geometrical figures, without which it is humanly impossible to understand a single word of it.* [6]

For the new question was not so much as how many angels could dance on the point of a needle, but rather how many could squeeze through the pinhole: for the new light was linear, clinically calculated and exclusive, and the spectrum of the earth and the rainbow of its people were about to enter a strange new territory, all rarefied crystal clear: thus to reveal a radical revision of the cosmos, new documentary and drama: literally the creation of a new heaven and earth announced by the cuckoo clock. For the angels were now knocked off their perches, their souls replaced by forces, all automated by the cuckoo above the Platonic arch.

Cuckoo said the clock, announcing the Deus ex Machina of the new dawn. For cuckoo it was, as time would tell. For Galilean physics and the great Newtonian machine to come may have worked, but actually, when it came down to it, made no sense: was inconceivable, quite absurd. Not that this occurred to Galileo, as far as he was concerned the rainbow was an illusion, of appearance only, residing in the mind of the perceiving subject, and as such is superfluous, a waste of time. For all the things of Beauty, of the Soul and its qualitative aspects, count for nothing, do not add up: for by virtue of their unique essence and quality, and in proportion to their immeasurable qualities they are as useless as their representations and signs, and provide no true knowledge of reality. For only bodies and things that can be measured are real, as so conceived in the object-light. Not that all these moral, spiritual and philosophical ramifications bothered the great man himself, who, without the luxury of retrospect, was quite oblivious and frankly couldn't careless, give a sweet white red or blue. And as he takes aim to toss his rocks into the Vatican and shatter the window of faith, he throws in a few choice words just for good measure, and gives it his two corporeal fingers with those now all too familiar words:

Now I say that when I conceive any material or corporeal substance, I immediately feel the need to think of it as bounded, as having this or that shape; as being large or small in relation to other things, and in some specific place at any given time; as being in motion or at rest; as touching or not touching some other body; and as being one in number or a few, or many. From these conditions I cannot separate such a substance by any stretch of the imagination. But that it must be white or red, bitter or sweet, noisy or silent, and of sweet or foul odour, my mind does not feel compelled to bring in as necessary accompaniments. Without the senses as our guides, reason or imagination unaided would probably never arrive at qualities such as these. Hence I think that tastes, odours, colours, and so on are no more than mere names so far as the subject is concerned, and they reside only in the consciousness. Hence if the living creatures were removed, all these qualities would be wiped away annihilated. [7]

Galileo is not going to just shatter those windows but annihilate them, thus wiping the Holy Spirit, and not least human consciousness, off the face of the earth. For The Logos had gone, had vanished, disappeared: right in front of everyone's eyes. Pity the Church which had been playing quite happily all along, and was actually having a lot of fun, knock it on the head, burn a few witches, lot of bottle Aristotle, nice shot. The game was up. For it had lost not only the touchstone of truth, but the entire stadium, whole premises in the process. For the future now lay with those traitors who not only failed to realise the deception, but actually were having a lot of fun playing the new game, all floodlit under the crystal palace. It was called Modern Science. And the Holy Fathers were left scratching their heads. For it mattered not whether the object was red, white, or blue, they could not even touch it, let alone thump it: it had gone. The Logos had vanished. Disappeared. Was daylight robbery to the biggest heist in the history of human consciousness, those crown jewels, pearls of soul, gems of wisdom, carefully enshrined in the Vatican. The City of God was being held to ransom. All courtesy of Galileo, who was getting on the Vatican's nerves. The Church was not amused: busy as it was fighting a rearguard action all over Europe, and wary upsetting the congregation and adding fuel to the reactionary flames, it found itself unable to resist Galileo, who was doing a dirty right on the doorstep, the clash of words and worlds coming through his infamous trial.

With Pythagoras, mathematical correlations were never material causes. It was symbiotic: different ratios gave different music, and vice versa, and they got along quite happily together. Ever since classical times the observation of the heavens had produced all sorts of ingenious geometries and epicycles to explain the rudimentary motion of the planets, which you could explain like music. It was a kind of cosmic game of algebra, and it was perfectly possible that two systems could explain the same motion and still play geometry together. Starting with Ptolemy the tradition arose whereby mathematical models to 'save the appearances' of planetary motion, and not speculate about real motion itself. Thus the Church, allied to a cosmology of Pythagorean harmonies and the music of the spheres, could tolerate a number of modest opinions, whilst retaining the right to have the final say in the matter. Hence its benign attitude, albeit wary and uncertain, to a timid Copernicus.

Galileo though was a tiger compared to Copernicus, was rigorous, methodical, solid and uncompromising, and his stocky red-haired matter-of-fact turned to remonstrate with Rome.

The litmus test had come, not that it mattered what colour the ball was or paper turned, whether red, green or blue, whether it pointed toward the earth or the sun – whichever whatway we colour the history – all are incidental to the surface of what was truly penetrate to and underlying the appearances: Galileo's uncompromising materialism and its annihilation of consciousness: ergo the soul. For infamous it may be, but something more was at stake than is generally understood in this epic battle. Something higher, wider and deeper than either the Church or Galileo or history or science would ever appreciate: for deep within the Logic of the Logos lay a curious twist of fate: the annihilation of consciousness of a planet and people living under an eclipse. For the crucifixion of nature had come, as exorcised of spirit. Galileo, who would never live to the results, had outdone himself, undone creation. And all because of the 'physical truth' that was at the nub of the problem. Question being, who the ruled of the universe? Who was at the centre – now was it the Pope or Galileo, earth or sun? Or had the question itself changed. Not so much who but – what? For it only took an atom — to split the difference, birth a new universe, and blow the bubble of the Medieval mind.

The Church, not to be bullied by a mere mortal, suggested to Galileo through the counsel of Cardinal Bellarmine, that it was quite permissible to discuss the Copernician system as a hypothesis to 'save the appearances', but not to assert it as a physical truth, and Father Grienberger, a bystander observing from within the Vatican, said that 'if Galileo had not incurred the displeasure of the company, he could have gone on writing freely about the notion of the earth to the end of his days.'

Bellarmine advised it would be unwise to push the issue, but Galileo ignored counsel, and pressed on in 1632 with his *Dialogue concerning the Two Great World Systems*. Written in Italian for a lay audience, the rhetorical ramrod swung into the doors of faith, it's echo resounding throughout Europe. The rock of ages, the Word and Logos, the Church and all Christendom, the Holy Spirit and all the Angels were in for a shock. With its spiritual authority, historical legacy, not least cosmology, made unacceptable to quantitative analysis and relegated to the ephemeral fringes of the fickle mind, there to contemplate in existential fancy, Rome — outwitted and rudely awakened from its slumber — reluctantly moved its ageing members, and found itself opportunely presented with the appropriate spectacle to confine the Galilean nuisance in its cosmological heresy. If Galileo ruled the heavens, then at least it ruled the earth, which happened to habitate Galileo, whose sin was the precocious certainty of his method, staked out on the door step of the Church. He was not accused of heresy but of arrogance. How dare he reduce God's creation, Joseph and Mary, the Blessed Virgin and all the angels and saints to mathematical notions, cut off the wings of faith, leaving the Vatican with only vans to beat the air.

Galileo was not remotely interested. What mattered was the mathematics. All else was irrelevant, distraction to task. Reality was not the hocus pocus qualities and potentials of Aristotle any more than it was the sympathies of ancients, nor the feelings and thoughts of the medieval, now subjective imagination. These are ephemeral. For Galileo had the master key to unlock the secrets of the universe. God had structured that universe according to mathematical principles, the way to understand it was by deducing these principles into fundamental laws in conjunction with observation and experiment. Want to know the Author? Then read the book of Nature. Open your eyes. See, observe divine principles at work. Deduce them from experiment. See God's divine hand moving around the earth like the sun, encompassed in the abstract light of reason. Heliocentrism was not the end but beginning of this quest. Neither man nor earth were at the centre of creation, nor really the sun, but something more distant, and more distant still, in the beyond of applied mathematics, as abstracted from reality by Galileo in the form of a modified Atomism.

The power of abstraction had arrived, and was about to sweep the old world away in the Herculean task of making over all unto a new likeness: the New Physics and its notions of mass, force and substance. Not that there was as yet a warrant for this assumption nor a foundation for its premise, and it would be left to Descartes to provide the philosophy and Newton the calculus. Meantime all Galileo was saying was that a sun centred system was mathematically commensurate, was more correlative and simple. Contrast this with Medieval fact and everyday intuition: for the sun revolving around the earth is as clear as sunrise. One only has to wake up to observe the sensibility of the Medieval mind, supported by the common sense notion that terrestrial objects are stationary, rooted to the earth, the natural place and focus of the universe. And if the earth did revolve, why didn't objects just spin off it, or if thrown up in the air, why did they not fall upon a different spot on the now – moving ground?

Only with Newton would these questions be answered. In the meantime Galileo persisted with his geometrical speculation of what was underlying the Medieval appearances. He followed the example of his ancient mentor, Archimedes, case in point being not the lever but the sun. Following the idea of deductive systematization, starting from first principles, the axioms of geometry, he devised his ideal, a clockwork sun, then tested his cosmic lever in the real world of terrestrial objects. And because the measurement of motion in this case was linked to the universe and at large, it would properly test the appearances, not save them. Such an approach appealed to the rough and ready, yet refined, turn of Galileo's mind. One critic captures the irony: 'Galileo typifies the direction of modern interests … not in refuting St Thomas, but in taking no notice of him. Motion might be all the angelic doctor declared it to be; Galileo nevertheless will drop weights from the top of the tower, and down inclined planes, to see how they behave.'[8]

In the form of a sun-centred astronomy, all else follows from first principles. Galileo assumes it as the most elegant solution and then sets about proving it, not only in the abstract but the real, in the application of his axioms to terrestrial objects, just as Archimedes had tested the principles of his lever in the real world. In Galileo's case though the whole cosmological order of the Church Scientific was held in the balance. And as physics became the touchstone of truth, so bridging the gap between the abstract and the real – the ideal and actual lever, the motion of the sun and the earth – came through empirical trial and tribulation, the new truths of the universe being sought in the commonplace, not the divine. Galileo's grand theory gravitated down to the behaviour of balls dropped from masts of ships, the measurements of tides and sunspots, the deductive consequences of observations playing trivial pursuit and pandemonium with the Medieval cosmos. For after Galileo, the Archimedes of the Scientific Revolution, nature would begin to tip over until she lost her balance, as everything was pushed over the edge, not least the great Church Apostolic, now dumped in a bucket of historical make-up. For Beauty had seen better days, was no longer the princess at the ball, now a mere shadow flung on the floor, all dazed and confused, rather frail, quite unsure, all those cracks in her make-up revealing love betrayed, together with the faint hint of hypocrisy of a desperate flirt, old wizened whore.

This metamorphosis, abrupt change from the spiritual to the material, the qualitative to the quantitative, the organic to the mechanical, occurred firstly in the mind, in the world of ideas, as Galileo's modified Atomism took hold upon the world, so as to become the new ideal, new Platonic universal unto which all particulars would apply. For all peoples and all things, all places and events, all things bright and beautiful, would be drawn into the process. And so the morality, meaning and purpose of the Medieval cosmos was kicked away and Beauty fell flat on her face. And whilst knocked drop-dead unconscious, looking out on a dark inchoate universe, wondering how she got there in the first place, so she came to analysing daisy chains, and chains and chains upon reason. As she slept the walls closed in. Closer and closer. What was above, was now down below, for hidden in those huge prominences, deep valleys and chasms of the moon was something now dead and inert, a skeleton, and many more skeletons to come, soon to rise in the rattle and rabble of mech-anism. Life had lost its meaning. For brushing away her petals, removing her pretty mask, there was betrayed an elaborate illu-sion of something more certain, more precise. And so Nature and Beauty were given the last rites, not so much through those two very corporeal fingers of Galileo, but through the exquisite genuflection of that famous Frenchman, the great high priest of the Matrix, René Descartes.

•

DESCARTES

Nowhere in the history of Western thought is the doctrine of mechanism so singularly combined as in the mind of René

Descartes, the founder of modern mechano-materialism. His contemporary Galileo may have been the Archimedes of the Scientific Revolution, but Descartes was the Platonic doctor, making complete the Greek metaphysic of Biblical revelation. He's about to turn Genesis into the Timaeus, going to finish where Aquinas left off. For now the Church was losing its grip, and *The Great Chain of Being* had broken at its weakest link, leaving clear a wide open path for a fully blown mechanical creationism embodied in the Medieval metaphor of a clockwork universe.

Not that we remember the cutting of the umbilical chord, nor do we even recognise the child as our own, but we do have a discourse on method and some meditations and principles on the procedures involved from the nurse practitioner, the Socratic midwife, as a memento of the occasion, all duly recorded by the great Frenchman himself, René Descartes, now lantern to our journey which comes in the form of an autobiography of a man born in search of truth.

After a thousand year slumber his generation is now in a period of rapid transition. Wealth, commercialism, state nation, the Renaissance, Reformation, everything was on the move as a new world was forged out of the old 'Middle Earth' of the Medieval mind. In this world Descartes is a philosophic Columbus embarked on a voyage of discovery, guided by a lantern to single-handedly establish the resettlement of the Medieval imagination upon new shores of truth. Like Columbus he is the guide across the great divide of faith and reason, constantly confronted by Middle Age ignorance. Unlike him, he conquers his territory solely within the sojourn of the mind, as body and soul are subjected to a method leading from mystical doubt to perfect clarity and certainty, forming a most peculiar combination of metaphysics and theology.

The original and major corpus of Descartes' journey is given in his *Discourse on Method* published in 1637, later to substantiated with *Meditations and Principles*. It is here that we join him alone in thought more than hundred years ago, as he describes the new maps, charts and bearings used to transport all souls on board across the great divide that now existed between faith and reason. Before we embark best be warned: as tortuous and methodical as the journey may be, we are led step by step to a simple end: for over and on the other side is the Shadow.

The *Discourse* was published in later life, when Descartes was forty one. Its full title reads *A Discourse on the Method of Rightly Conducting the Reason, and Seeking Truth in the Sciences.* Candid, like its nomenclature, written in French rather than ecclesiastical Latin, it begins without Medieval fuss or ornament.

> *Good sense is of all things amongst men, the most equally distributed; for everyone thinks himself so abundantly provided with it ...* [9]

Henceforth the author and prophet of Mechanism is about to redefine what exactly 'good sense' is. First though comes doubt. Doubt of accepted learning, doubt of knowing and reason, of the senses and mind, before a final certainty and synthesis of knowledge is reached. This certainty comes stage by stage, divided into the six parts of the Discourse, the first half forming a preliminary parting of the waves before the final showdown.

In the first three parts Descartes sets sail, disillusioned with Middle Age ignorance. He finds 'the varied course and pursuits of mankind at large ... as vain and useless', but if there is any one worthy of pursuit then it is the one he has chosen – the culture of his reason. After studying in 'one of the most celebrated schools in Europe', and having 'read all the books that had fallen into my hands', he decides there is 'no science in existence that was of such a nature as I had previously been given to believe', and 'entirely abandoned the study of letters ... spending the remainder of my youth travelling, in visiting courts and armies.' Two years later in 1619, whilst serving as a mercenary for the Duke of Bavaria, he experiences the most momentous day of his life.

'Being full of enthusiasm and 'having discovered the basis of a new science', he is possessed of a vision to 'wholly sweep away' all he has learned and devote himself to discovering a new world. Ten years later he travels to Holland to live a life as 'solitary and retired as in the midst of the most remote deserts', where he decides to put up with Middle Age ignorance, 'expediency seemed to dictate that I should regulate my practice conformably to the opinions of whom I should have to live', advises some of his followers to stay at home, 'the single design to strip oneself of all past beliefs is one that ought not be undertaken by everyone', before his thoughts become caught up in the present intensity and discovery of the new world of Part Four:

> *I am in doubt as to the propriety of making my first meditations ... for these are so metaphysical, and so uncommon, as not, perhaps, to be acceptable by everyone ...* [10]

Descartes is at the doorway of a strange new world *different in kind* to anything before, is the first person, adam of the cave, to be birthed into a new vision of the cosmos. So far no one has entered, except for Galileo, who had dropped a few balls into the new well of consciousness, and encouraging indeed was the effect: nice and uniform was the echo, and very orderly it was too, now cause to applause resounding throughout Europe. Paper and pencil at hand with water-clock, it was all rather rudimentary stuff, destined one day to take man up to the heavenly sphere of the Aristotelian pearl of the moon, thus to leave an imprint on Galileo's Messenger of the Stars. This step is now taken to its ultimate conclusion by Descartes who is about to take us into the abyss, as he extends Galileo's philosophy to the deep valleys and dark chasms of the human soul, now a shadow of its former self, in the great culmination of Greek and Christian thought.

It's a step in the dark, a leap unto the abyss, into an error that possibly one day – and then perhaps too late – would be recognised for what it was. For, considering the regulations and stipulations set forth by Galileo, this is going to be very dark and strange place indeed, and whoever goes under the arch and over the cuckoo clock is going where no man has gone before. For Galilean revolution may have taken man to the pearly gates of heaven, the physics was alright, the problem was the rest of it, that rather large expanse called the mind, which is rather more metaphysical and differently inclined, actually home to consciousness, now left in poverty, defined out of the premises. For when it came to where it mattered, in the realities of the human heart and soul, for all its rocket science the Galilean odyssey was a vanity of balls, a fleeting shadow. The leaning tower of Pizza might have been a convenient experimental apparatus for dropping things, and it was all very well for Galileo to abstract his philosophical notions for the behaviour of balls, but what about man – was he that way inclined? What happened when you dropped him in it? Bit different from balls down inclined planes, and cannon shot lobbed from masts. Or was he cannon fodder too? For no one had as yet tried that experiment. Abstract notions may be all very fine for balls, and no one was particularly bothered what colour Galileo painted them, they still behaved the same, and who could blame him for removing such unnecessary and superfluous accompaniments. Such was his no-nonsense approach. All very fine for going about your business, especially if you're a mathematician. But what happens when you get carried away and take a step too far – that leap in the dark?

For as yet no one had put their hand through, let alone stuck their neck out under the cuckoo clock to peer through the Platonic arch, the entrance to the cave. For did the hand, having no qualities as such, simply vanish, disappear? And who would go there? Dare venture into the isolation chamber of those primary quantities, a place where there is only number and measure, and little else. What was it like? And who was up for it, this tortuous logic, this experiment in the crucible? It's not something in which you can, like Galileo, drop a few balls and scribble some algebra, cook up an experiment, juggle recipes and make dinner. For what is on the plate is you. You can't just sit there, figure it out and exclaim voila, c'est ca. You have to go through it. It's a state of mind, different in kind, and as such it's going to take some serious meditation, mental gymnastics to remove all that spirit, narrow the doors of perception, create that vacuum, and squeeze yourself through the door – or rather the pinhole – of Plato's brand new academy. Cue Descartes, as true and pious a Catholic as there ever was one: he's going to save the Aristotelian cosmos, and oblige Galileo by going very modern: for he is about to discover himself and do some meditation:

> *I will close my eyes, I will stop my ears, I will turn away my senses from their objects, I will even efface from my consciousness all the images of corporeal things; or at least, because this can hardly be accomplished, I will consider them empty and false; and thus, holding converse only with myself, and closely examining my nature, I will endeavour to obtain by degrees a more intimate and familiar knowledge of myself...*[11]

In his solitude, Descartes asking himself the big question – or rather in French, he's endeavouring to obtain by degrees a more intimate and familiar knowledge of himself. However that's as far as the meditation goes, for this is no Buddhist retreat, but a metaphysical workshop, and very modern at that. As Descartes enters our cubicle and sits on the chair, he's come prepared with his *Meditations and Principles*, paying capital attention to the 'I' of what is going to be a very modern ego.

> *It was absolutely necessary that I, who thus thought, should be somewhat; and I observed this truth, I think hence I am, was so certain and of such evidence, that no ground of doubt, however*

extravagant, could be alleged by the sceptics capable of shaking it, I concluded that I might, without scruple, accept it as the first principle of the philosophy of which I was in search.

In the next place, I attentively examined what I was, and I observed that I could suppose that I had no body, and that there was no world nor any place in which I might be; but that I could not therefore suppose that I was not; and that, on the contrary, from the very circumstance that I thought to doubt of the truth of other things, it most clearly and certainly followed that I was; while on the other hand, if I had only ceased to think, although all the other objects which I had ever imagined had been in reality existent, I would have had no reason to believe that I existed; I thence concluded that I was a substance whose whole essence or nature consists only in thinking, and which, that it may exist, has need of no place, nor is dependent on any material thing; so that "I", that is to say, the mind by which I am what I am, is wholly distinct from the body, and is even more easily known than the latter, and is such, that although the latter were not, it would still continue to be all that is. [12]

Descartes has entered our cubicle, the crucible. He's closed his eyes, closed the door and he's not here to pray, nor has he come to contemplate or meditate like an Eastern sage or Christian mystic: in fact his intentions, right from the outset, are quite the opposite: he is only concerned with himself, the 'I', the object of thought, his ego: for only thought gives ground to his being, turns him into an object of self, defines him against the background of being, as some 'body' or 'thing'. What Galileo does to bodies, 'now I say that when I conceive any material or corporeal substance, I immediately feel the need to think of it as bounded', so Descartes does to consciousness – now the 'shadow of thought', because Descartes is absolutely convinced that his self existence is so conceived by thought: and his ego has just cause, just like Galileo's, because thought, as construct of mind, does work like a shadow: it separates the subject from the object, divides them into two, into 'this' and 'that' – is the very definition that self consciousness, the sense of separation, the very divide, split, that not just mystics, but we all everyday intuit and transcend – through the touchstone of feeling. But Descartes is having none of it, and as he comes to analyse the nature of what it means to 'think' he does so closely and methodically, offering a definition, that is, most assuredly, one of the most daring and precise as there has ever been. He then uses it to establish it as the very definition of his identity as a separate being akin to Galileo, as a body, a 'thinking thing', 'bound' in some sense, as he defends his unshakeable conviction.

This 'I' of Descartes, this 'me' so self conceived is now the shadow. As Descartes cuts the head from the heart, we find mind in mid-ocean, in some sort of disembodied state, where locality and sensibility have gone, as body unto shadow, a shadow aware of one indubitable certainty: that it is an object of thinking, object-thought, a uniform 'substance whose whole essence of nature consists only in 'thinking', and this shadow, divested of body and consciousness, all spiritual qualities, abstracted of all secondary properties, is 'wholly distinct from body' and is not only more easily known, but in some sense 'all that is': immortal. But what is the nature of this thought substance, this 'thinking thing', that endures beyond the body? As to the nature of the shadow Descartes attends to in his next paragraph.

After this I inquired in general into what is essential to the truth and certainty of a proposition; for since I had discovered one of which I knew to be true, I thought I must likewise be able to discover the ground of this certitude. And as I observed in the words a I THINK, HENCE I AM, there is nothing at all which gives me assurance of their truth beyond this, that I see very clearly that in order to think it is necessary to exist, I concluded that I might take, as a general rule, the principle,

*that all things which we clearly and distinctly conceive are true,
only observing, however, that there is some difficulty in rightly
determining the objects which we distinctly perceive.* [13]

I think therefore I am.
Says the Shadow,
now disembodied,
now a 'thinking thing' cast off,
wholly separate,
as an object of thought.

But if this shadow is wholly distinct from the body and the
world, what exactly is it and where's the connection? How is
Descartes going to bridge the great divide, the abyss that now
exists between he and the shadow? How is 'It' going to act? It's
stuck, for it has no general rule or principle in this new found
territory of the soul. This shadow, this thinking thing cannot
appeal to the phenomenal world of the senses, anything apart
from the indubitable certainty of thought. According to Descartes the shadow must be made 'clear and distinct' in order to
find a method for rightly conducting itself. For these clear and
distinct ideas must originate from somewhere that is apart from
consciousness, and lead somewhere that is not merely made up
of his subjective imaginings, they must be 'external' in some
sense: for the shadow must have an object, be a thing.

He invokes this external causation of the shadow through
the only option now open to him: by proxy of God, now the
divine Cause. The gulf between the self and the shadow, self
and world, in now resolved through the prism of the Cartesian mind: for God is now logical substance acting in him and
through the world. For nature is not now likened to soul or
consciousness, or any spiritual corpus of truth, rather it is anal-
ogous to the timeless purity of mathematical relations, in a logical perfection prescient in those clear and distinct relations of
a triangle:

> *supposing a triangle to be given, I distinctly perceived that its three angles were necessarily equal to two right angles, but I did not on that account perceive anything which could assure me that any triangle existed: while on the contrary, recurring to the examination of a Perfect Being, I found that the existence of the Being was comprised in the ideas in the same way that the equality of its three angles to two right angles is comprised in the idea of a triangle ... and that consequently it is at least as certain that God, who is this Perfect Being, is, or exists, as any demonstration of geometry can be ... whence it follows that our ideas of notions, which to the extent of their clearness and distinctness are real, and proceed from God, must to that extent be true.* [14]

Thus the First Cause of creation has been made complete through Descartes's 'substance' behind the shadow, now the necessary cause of all in the new belief system, catechism of existence: and the adam of the beginning comes to be not by the genesis of the Logos but rather the timaeus of Logic, for the Greek Demiurge has gone Christian, as exemplified by Descartes: for what is clearly and distinctly conceived and comprehended in the object of speculative geometry, really exists external to me. And so the Bible has a new genesis, and Christianity a new creation, through the deification of science. The Logic, beginning with Aquinas, had reached its final consummation, and the new creature emerging from the Galilean well of consciousness is that strange metamorphosis called modern man, who puts his foot right in it, to trip over the premises, in the great –

I think therefore I am {GOD}.

God now hidden in the premises by parenthesis, God now the logical necessity of the Matrix. As Descartes, the new high priest, holds up object on parapet, so the high mass is performed, consubstantial with the Shadow. For in it, with it, and through it we are made – in the genuflection of the triangle.

As we stare unto the abyss, Descartes hasn't blinked an eye and is already on parts Five and Six, where all he needs to do is apply his new found principles to the world by observing 'those laws established by God in such a manner that ... we cannot doubt that they are accurately observed in all that exists or takes place in the world.' And the genesis of this new age is now described by Descartes through an Atomistic account of a Greek Demiurge, resolved as he is 'to leave all the old people to their disputes, and to speak only of what would happen in a new world, if God were now to create somewhere in the imaginary spaces matter sufficient to compose one, and were to agitate variously and confusedly the different parts of this matter, so that there resulted a chaos as disordered as the poets ever feigned, and after that did nothing more than lend his ordinary concurrence to nature, and allow her to act in accordance with the laws he had established. And these laws are universal, for 'even if god had created more worlds, there could have been none in which these laws were not observed.'

For everything now is to be observed under the object-light and, 'Present knowledge is almost nothing in comparison to what remains to be discovered.' Examples follow as Descartes advises us to 'take the trouble of getting dissected in their presence the heart of some large animal', which works 'according to the rules of mechanics which are the same with those of nature.' Brain changes produce 'waking, sleep and dreams ... light, sounds, odours, tastes, heat and all the other qualities of external objects', and this mechanical account will not 'appear at all strange to those who are acquainted with the variety of movements performed by different automata, or moving machines fabricated by human industry', and, 'such persons will look upon this body as a machine made by the hands of god, which is incomparably better arranged, and adequate to movements than is any machine of human invention.'

Minus doubt plus certainty, pertaining to a reality now distinctly perceived, calculating clear, perfectly provident, save for Pythagoras or even Plato, Descartes is perhaps the greatest philosopher that ever lived: lest we forget, he is the true author of modern materialism, the prophet of all that was to come. For he had the courage to go where no man had gone before, and through him was made a new metaphor, an analogy that would prove spectacularly successful, yet so fatally flawed, as the deified cause becomes an end in itself: the end of nature. Finally let Descartes himself conclude his Meditations, edited as epitaph:

> *For I perceived it to be possible to arrive at knowledge highly useful in life; and in the room of speculative philosophy usually taught in the schools, to discover a method, by means of which, knowing the force and action of fire, water, air, the stars, the heavens, and all other bodies that surround us, thus render ourselves the lords and possessors of nature.* [15]

All that was lacking was a calculus to make the shadow complete, so as to prove the one-to-one correspondence. Come the lord of the force of earth, air, fire and water, come the true possessor of the new Science of Nature. Come the Divine Engineer, Sir Isaac Newton.

•

NEWTON

Speculation is useless without proof, and proof came with Newton whose grand synthesis started where Descartes left off. For Descartes may have dangled his deified precepts upon a triangle, and the logic may have led to the Shadow, but a shadow will not fall without a wall, some substance to reflect upon, so that it may be shown to work, proven to act, as measured by a perfect calculus, a one-to-one correspondence with the object in question. Come the man who built the Wall, the man of the moment, the man who, in search of the corporeal substance of light, literally looked straight into it and got blinded by the sun, then spent three days recovering in his room; the man who carried an ever present Bible by his side; the man who did not feign hypothesis, and finished off the demolition process: for if Mechanism proposed a one-to-one correspondence with reality just as in geometry, then where exactly did God fit in? The answer was nowhere, but took a little time forthcoming.

Each successful mapping of the New Science onto nature revealed that behind the divine mystery of creation lay a rational explanation, and as topography advanced so the spirit retreated into smaller and smaller reservations. As squeezed and probed by the New Philosophy God found himself cornered and summarily requisitioned and made over to His proper likeness in correspondence to with Shadow. The premises of the cave, beginning with Aquinas, now reached their conclusion as mediated by Mechanism. He was out of the picture for good.

Originally Newton, like Descartes, like Galileo, like all the Mechanical Philosophers of the seventeenth century, saw Mechanism as an essential and timely support for religion just as its Aristotelian foundations were crumbling. Its new truths may have been a little suspect, but it seemed fair game. Powerful and precise, prudent and provident, regular and reliable – the new God was a true god, the sort of god who always had the answers, never let you down. So places filled for the pilgrimage to the new world. The possibilities seemed endless. Once cast off Medieval shores essentially there was no return – the bridges were burnt – to a world different in kind from anything hitherto conceived by the human mind. Descartes had done worse than de-deify the Church, rather he had used it as vehicle to support his own vision of the cosmos. Spring cleaned and dusted of all unnecessary furniture and superfluous ornaments, the Newtonian cosmos to be was a spotless home, a Platonic Republic based on a solid logic to replace Rome, now the perfect abode, purchase upon fate, prudently insured against defects – defects now erased, defects that came up dazzling clean in the washing machine, the Platonic heaven of essentially pure mathematical relations. Unsurprisingly from the Timaeus came Deism, then scepticism and finally atheism as the profundity of modern thought became inversely proportioned to its profanity of spirit, removing those necessary barriers to conquest out the way. Or, in the words than Stuart Hall:

> It was in a sense a bargain, attested by Descartes, that nature and the animals should be handed over to mechanistic science while God and the soul and all the ultimates rested with philosophers and the divines … but just as the autonomy of science gave it freedom to exploit successfully a materialistic metaphysic within the harmlessness of its own confined domain of natural phenomenon, so the very connections of science and religion ensured that, once established, this metaphysic must diffuse beyond the postulates of scientific enquiry. For these limits were the products of an illusion, one to which the greater and minor scientists from Galileo to Newton had subscribed, no less than the philosophers and men of religion who sooner or later embraced and applauded their ideas. The former were deluded into imagining there is in the nature of things a barrier to the materialist metaphysic which men could never be eager or able to cross, a barrier for ever protecting the psychic and spiritual realm. [16]

Newton and his generation shrank back from crossing this barrier because it would establish a new religion, state and independence, implying radical changes in the human constitution not even they were willing to contemplate. As usual the enemy within was the *Logic of the Logos* as the contradictions of mixing a Biblical God with Greek geometry reached their final solution in *Newton's Principia* and the rarefied heights of the calculus. Published in 1687, *The Mathematical Principles of Natural Philosophy*, (now heralded as the single most important tract in the history of science), was so far ahead of its time, that only a handful of Newton's contemporaries could follow it, and even they were confused. Its makings were touched with typical Newtonian idiosyncrasy.

Twenty years before, Newton had been moved by the eccentric behaviour of the moon to roughly calculate the motion of the planets, assuming gravitation as the basis of Kepler's Third Law. But he had better things to do than 'feign hypothesis', and whilst preoccupied with mathematics and optics, grinding his own lenses and staring into the sun, dabbling in alchemy and Biblical criticism, customarily taking thousands of notes, the crowning achievement of his life was left to fester. Only in middle age and when visited at Cambridge by the famous astronomer Halley, did the mathematical mastermind oblige an answer from his Laucasian chair. It wasn't even his special subject, and the little cosmic revision done in his youth was unbeknown to his unsuspecting audience. The big question seemed innocuous enough, as put by Halley: "What would be the curve described by the planets on the supposition that gravity diminished as the square of the distance?"

Newton immediately answered: *An Ellipse.* Struck with joy and amazement, Halley asked him how he knew it? Why, replied he, I have calculated it; and being asked for the calculation, he could not find it, but promised to send it to him.[17]

Careless Isaac fumbled for the formula that had eluded his contemporaries. On examination of his notes, with typical penchant for perfection, Newton saw they were in need of comprehensive revision, out of which emerged the *Principia.* Describing the ellipse and its complex relations to the solar system as a whole called for a new mathematics to compute the force of all matter in motion. Encapsulating a complete physics, coordinated in Absolute Time and Space, accompanied by twenty five pages of Euclidean axioms and definitions, the modern universe was now framed by a universal calculus of such power and precision that its one-to-one correspondence to reality would have universal applications. From the astronomic proportions of the universe to the atomic configurations of matter, the fine tuned calculus was to apply: all framed upon the solid Newtonian Wall of Absolute Space and Time.

Let us watch the momentous occasion.
Newton is up on the parapet by the tree.
The apple falls. The shadow is observed.
And the calculus is conferred by all asunder,
all in perfect symmetry, consubstantial wonder:
a perfect one-to-one correspondence to the shadow,
as beheld in the object-light.

Nature and nature's laws lay hid in night.
But God said, "Let Newton be, and there was light."

So declared Pope Alexander. Morning had broken, the apple had fallen, Newton had spoken, and the Matrix was born. Trouble was even Newton scratched his head, for the mathematical matrix and its whole rational superstructure was an enigma to the very mind that had created it. All he did was the sums: the rest was none of his business, nor did he really care to commentate. For he did not 'feign hypothesis' as to the real nature of things. He just did the calculations. Mesmerizing as the calculus was in its shadow correspondence to bodies, was it a cause? When Galileo measured the behaviour of balls down an inclined plane and gave formulae to motion, were these 'causes'. When Pythagoras measured the ratio of harmonies on a lyre, were these 'causes'? When we read a music manuscript, the notes that call the tune, are these 'causes'? They might be signs, might be correspondences, might be processes. But causes? For Gravity itself, as Newton well knew, was a mathematical fiction, an 'absurdity', quite inconsistent with the precepts of Mechanism. In his own words:

> *That Gravity should be innate, inherent and essential to Matter, so that one Body may act upon another at a Distance thro' a Vacuum, without the Mediation of anything else, by and which their Action and Force may be conveyed from one to another, is to me so great an Absurdity, that I believe no Man who has in philosophical Matters a competent Faculty of thinking, can ever fall into it.* [18]

When the eighteenth century materialists talked blandly of the great clockwork universe and the Divine Engineer behind it, little did they realise the force behind it was 'occult' and acted at distance, that Gravity itself was lost in translation, was a 'thing in itself', as Kant might say, something you may measure, but never truly define – just like all the atoms and quarks to come that were different in kind, things to replace the angels as the prime movers of the universe, things that were shadows, mathematical fictions: for there was no such 'thing' as Gravity any more than there is a 'Graviton'. Rather is was an ideal concept, instrumental notation, a metaphysical construct, abstract notion, contingent fact: a shadow on the wall. Not that it mattered. The simple fact was it worked, and the calculus was the key that opened the door to new universe, opening up the world of large and small, with many corridors aside, as other departments followed suit in the wake of Newton's vision of a grand synthesis of knowledge to to unite the forces of science, as speculated in his *Principia*.

> *I wish we could derive the rest of phenomena of Nature by the same kind of reasoning from mechanical principles; for I am induced by many reasons to suspect that they may all depend*

upon certain forces by which the particles of bodies, by some causes hitherto unknown, are either mutually impelled towards each other, and cohere in regular figures, or are repelled and recede from each other. These forces being unknown, philosophers have hitherto attempted the search of Nature in vain; but I hope the principles here laid down will afford some light either to this or some truer method of philosophy. [19]

As to the aforementioned particles, Newton subscribed to a modified Atomism, as appended in the General Scholium, added to the Principia in 1713:

It seems probable to me that God in the Beginning form'd matter in solid, massy, hard, impenetrable, movable Particles, of such Sizes and Figures, and with such other properties and in such Proportions to Space, as most conduced to the End for which He form'd them. [20]

Newton, much like Descartes, is thoroughly Greek, and was worshipped, adulated, as first lord and possessor, if not reluctant member, of the admiralty of science as it heralded the new age of the machine. Made and set in motion by God such a machine was precise: nothing happens by chance, nothing is arbitrary, for in reality all things are traceable to the fundamental laws of nature. Eternal, infinite, self moving, such a machine was to prove not only the central problem of eighteenth century understanding but a source of perplexion for a long time to come: for how could a shadow cause an object? According to Leibniz, Newton could not have both God and Machine, both Mechanism and his Bible, and when challenged, Newton kept an uncomfortable silence. Instead fate was forced to choose, as the God of Newton, relegated to presiding over a mechanical creation, keeping it in motion and fixing it occasionally, proved to be incompatible with his philosophy. And as the shadow moved over and the light rescinded so God passed away: out of sight, out of mind. Anything other was a act betrayal of the new spirit of the times: for nothing could pervert the course of justice, now the calculus, law of the universe. It was just the way it was, and man like God, was now mere matter in motion, brought about by Accident and Chance, as fashioned by a modified Atomism. In other words, nature was now:

> a dull affair, soundless, scentless, colourless; merely the hurrying of material, endlessly, meaninglessly. However you disguise it, this is the practical outcome of the characteristic scientific philosophy which closed the seventeenth century. No alternative system of organising the pursuit of scientific truth has been suggested. It is not only reigning, but it is without rival. And yet - it is quite unbelievable. [21]

Being a shadow, God cannot be God. Or he was another sort of god, God of the Shadow. So much was obvious to the muddled logic of those 'realists' who were about to confer it upon creation. And as the religious scaffolding was finally drawn away the edifice did not fall down, as Newton feared, but was left fantastic, enigmatic, by itself: now the wall, as seen in a dark mirror of a world that would never to see the light of day, could never begin to conceive, let alone yet contemplate, the dark duplicity in which it found itself. Best leave it to Milton gives a flavour of what was in the wind:

> *The spirit of man, no longer confined within the dark prison house, will reach far out and wide till it fills the whole world and spaces far beyond the expansion of its divine greatness. Then at last, most of the chances and changes of the world will be so quickly perceived to him who holds this stranglehold of wisdom that hardly anything in his life is unforeseen or fortuitous. He will seem to be one whose rule and dominion the stars obey, to whose command the earth and sea hearken, and whom*

the winds and tempests serve; to whom lastly, mother nature herself has surrendered as if indeed some God had abdicated the throne of the world and entrusted its rights, laws, and administration as governor. [22]

After all, the Scientific Revolution had not ended. It had just begun. Newton had spoken. The apple had fallen. But not just the apple but the tree, felled by the axe of an infallible Logic, the fall now inevitable ... like gravity a foregone conclusion, the fall of man, the fall of creation, the fall of a planet into the abyss. Trouble was – that falling – no one saw it coming, had the faintest idea, heard it downing, hidden as it was in the dark forest, the invisible premises, of the most holy of holys of modern science. For the eye had not seen, no ear had heard, no mind had conceived, what was coming: the day of judgement, the final reckoning of the shadow, that most shallow and superficial of creeds: the modified Atomism of a Western materialism. For the day was coming, and the earth was moving, toward the eclipse, to Totality, the end of the great illusion that was the Matrix: to thus give way, so as to come to pass, to sunlight, the hope of humanity resting, believing, dreaming of a new horizon, the new dawn, the great awakening that must come, if truth be truth, no object in the way, nothing upheld, nothing stopping, the free passage of Light – of Love.

• • •

It's on the way – **The Second Contact.**

No better words than Pascal:

*The eternal silence of these infinite spaces fills me with dread,
I look in every direction and all I see is darkness … we run
heedlessly into the abyss after putting something in front of
us to stop us seeing it.*

So we move premises

run heedlessly into the abyss

after putting something in front of us >

                          to stop us seeing it.

The Eclipse
of the modern mind

A TRILOGY

Second Contact
on the way

●

References

Plato's Allegory of the Cave:
Plato *The Republic* (Penguin 1973) pp. 278-281

1: Prelude
1. John Losee *An Historical Introduction to the Philosophy of Science* (Oxford 1988) p.17
2. D. Pepper *The Roots of Modern Environmentalism* (Croom Helm 1984) p.49
3. Sir James Jeans *The Mysterious Universe* (Cambridge University Press 1931) p.137
4. Stephen Hawking *A Brief History of Time* (Bantam Books 1996) p.193
5. Professor Brian Cox, Andrew Cohen *Human Universe* (Collins 2014) p.6

2: The Greeks
1. Bertrand Russell *The History of Western Philosophy* (Unwin 1979) p.60
2. Bertrand Russell *The History of Western Philosophy* (Unwin 1979) p.62
3. Bertrand Russell *The History of Western Philosophy* (Unwin 1979) p.63
4. Bertrand Russell *The History of Western Philosophy* (Unwin 1979) p.63
5. Bertrand Russell *The History of Western Philosophy* (Unwin 1979) p.62
6. W.K.C Guthrie *A History of Greek Philosophy* Volume 2 (Cambridge University Press, 1962) p.39
7. Bertrand Russell *The History of Western Philosophy* (Unwin 1979) p.85-86
8. W.K.C Guthrie *A History of Greek Philosophy* Volume 2 (Cambridge University Press, 1962) p.28
9. John Losee *An Historical Introduction to the Philosophy of Science* (Oxford 1988) p.28
10. Bertrand Russell *The History of Western Philosophy* (Unwin 1979) p.56
11. Bertrand Russell *The History of Western Philosophy* (Unwin 1979) p.158
12. Bertrand Russell *The History of Western Philosophy* (Unwin 1979) p.140

3: Revolution in Science

1. Richard Tarnas *The Passion of the Western Mind* (Pimlico 1995) p.114

2. Richard Olson *Science Deified and Science Defied* (Berkeley UCP 1982) p.127

3. Arthur Koestler *The Sleepwalkers* (Penguin 1986) p.264

4. Arthur Koestler *The Sleepwalkers* p.345

5. Arthur Koestler *The Sleepwalkers* p.261

6. John Losee *An Historical Introduction to the Philosophy of Science* (Oxford 1988) p.17

7. D. Pepper *The Roots of Modern Environmentalism* (Croom Helm 1984) p.49

8. Basil Willey *The Seventeenth-Century Background* (ARK Paperbacks 1986) p.22

9. René Descartes *A Discourse on Method, Meditations and Principles* (Everyman's Library 1984) p.3

10. René Descartes *A Discourse on Method, Meditations and Principles* p.26

11. René Descartes *A Discourse on Method, Meditations and Principles* p.95

12. René Descartes *A Discourse on Method, Meditations and Principles* pp.26-27

13. René Descartes *A Discourse on Method, Meditations and Principles* p.27

14. René Descartes *A Discourse on Method, Meditations and Principles* pp.29-30

15. René Descartes *A Discourse on Method, Meditations and Principles* p.49

16. A.Rupert Hall *From Galileo to Newton* (Dover 1981) p.340

17. A.Rupert Hall *From Galileo to Newton* p. 293

18. A.Rupert Hall *From Galileo to Newton* p.314-15

19. A.Rupert Hall *From Galileo to Newton* pp.244-245

20. A.Rupert Hall *From Galileo to Newton* pp.237-238

21. A.N. Whitehead *Science and the Modern World* (Cambridge University Press 2011) p.69

22. D. Wolfe (editor) *The Complete Prose Works of John Milton* (Yale University Press 1939) p.98

BIBLIOGRAPHY

SELECT BIBLIOGRAPHY - *HIGHLIGHTS ARE RECOMMENDED READS.*

ALLAN DJ THE PHILOSOPHY OF ARISTOTLE OXFORD (UNIVERSITY PRESS 1970)

APPLEYARD, BRYAN *UNDERSTANDING THE PRESENT* (PICADOR 1992)

ASHVAGHOSHA *THE AWAKENING OF FAITH* (DOVER PUBLICATIONS 2003)

BARFIELD, OWEN *SAVING THE APPEARANCES* (WESLEYAN UNIVERSITY PRESS 1988)

BECKER, ERNEST *THE DENIAL OF DEATH* (THE FREE PRESS 1973)

BLAKE, WILLIAM *POEMS AND LETTERS* (PENGUIN BOOKS 1986)

BOHM, DAVID *WHOLENESS AND THE IMPLICATE ORDER* (ROUTLEDGE & KEGAN PAUL 1980)

BOHM, DAVID *CAUSALITY AND CHANCE IN MODERN PHYSICS* (ROUTLEDGE & KEGAN PAUL 1984)

BRAUDEL, FERNAND *ON HISTORY* (WEIDENFIELD & NICOLSON 1980)

BRONOWSKI, JACOB *THE ASCENT OF MAN* (BBC 1976)

BRONOWSKI, JACOB *SCIENCE AND HUMAN VALUES* (HARPER AND ROW 1965)

BRONOWSKI, J. *WILLIAM BLAKE* (PENGUIN 1984)

BROWN, HANBURY *THE WISDOM OF SCIENCE* (CAMBRIDGE 1986)

BUCHDAHL, G. *METAPHYSICS AND THE PHILOSOPHY OF SCIENCE* (BLACKWELL 1969)

BUCKE, RICHARD *COSMIC CONSCIOUSNESS* (NEW YORK: DUTTON 1969)

BURTT, E.A. *THE METAPHYSICAL FOUNDATIONS OF MODERN SCIENCE* (DOVER **2003**)

BUTTERFIELD, H. *THE ORIGINS OF MODERN SCIENCE* (LONDON 1949)

DE CHARDIN, TEILHARD *THE PHENOMENON OF MAN* (HARPERCOLLINS 1971)

CAPRA, FRITJOP *THE TAO OF PHYSICS* (FLAMINGO **1983**)

CAPRA, FRITJOP *THE TURNING POINT* (WILDWOOD HOUSE 1982)

CARSON, RACHEL *SILENT SPRING* (PENGUIN 2000)

COPLESTON, FREDERICK *HISTORY OF PHILOSOPHY VOLUME 1: GREECE AND ROME* (A&C BLACK 2003)

CUPITT, DON *THE SEA OF FAITH* (SCM 2010)

CUPITT, DON *THE LEAP OF REASON* (SHELDON PRESS 1986)

DAWKINS, RICHARD *THE BLIND WATCHMAKER* (PENGUIN 1990)

DESCARTES, RENÉ *A DISCOURSE ON METHOD, MEDITATIONS AND PRINCIPLES* (EVERYMAN'S LIBRARY 1984)

DIJKSTERHUIS, E.J. *THE MECHANIZATION OF THE WORLD PICTURE* (CLARENDON **1961**)

DUHEM, PIERRE *TO SAVE THE PHENOMENA* (UNIVERSITY OF CHICAGO PRESS 1969)

FEYERBAND, PAUL *AGAINST METHOD* (VERSO 1978)

FROMM, ERICH *TO HAVE OR TO BE?* (HARPER AND ROW 1976)

FUKUYAMA, FRANCIS *THE END OF HISTORY AND THE LAST MAN* (HAMISH HAMILTON 1992)

GALBRAITH, JOHN KENNETH *THE AFFLUENT SOCIETY* (PELICAN 1971)

GRAVES, ROBERT *THE GREEK MYTHS* (PENGUIN 1975)

GRIFFITHS, JAY *PIP PIP* (FLAMINGO **1999**)

GUTHRIE, W.K.C *A HISTORY OF GREEK PHILOSOPHY VOLUMES 1&2* (CAMBRIDGE UNIVERSITY PRESS, **1969**).

GUTHRIE, W.K.C *THE GREEK PHILOSOPHERS: FROM THALES TO ARISTOTLE* (ROUTLEDGE **1969**)

HAWKING, STEPHEN *A BRIEF HISTORY OF TIME: FROM THE BIG BANG TO BLACK HOLES* (BANTAM BOOKS 1996)

HEISENBERG, W. *PHYSICS AND PHILOSOPHY* (PENGUIN BOOKS 1979)

HOBBES *LEVIATHAN* (EVERYMAN 1973)

HOFSTADER, DOUGLAS *GODEL, ESCHER, BACH: AN ETERNAL GOLDEN BRAID* (PENGUIN 1980)

HOOYKAAS, R. *RELIGION AND THE RISE OF MODERN SCIENCE* (SCOTTISH ACADEMIC PRESS 1973)

HOWARD, JONATHAN *DARWIN* (OXFORD 1982)

HUME, DAVID *ENQUIRIES CONCERNING HUMAN UNDERSTANDING AND THE CONCERNING THE PRINCIPLES OF MORALS* (CLARENDON PRESS 1975)

JAMES, WILLIAM *THE VARIETIES OF RELIGIOUS EXPERIENCE* (PENGUIN BOOKS 1983)

JEANS, SIR JAMES *THE MYSTERIOUS UNIVERSE* (CAMBRIDGE UNIVERSITY PRESS 1931)

JESSOP T.E. *BERKELEY PHILOSOPHICAL WRITINGS* (NELSON 1952)

JONES, ERNEST *THE LIFE AND WORK OF SIGMUND FREUD* (PELICAN 1964)

JONES, ROGER *PHYSICS AS METAPHOR* ABACUS 1983

JUNG, KARL *MEMORIES, DREAMS, REFLLECTIONS* (PENGUIN 1979)

KIERKEGARD, SÖREN *FEAR AND TREMBLING* (PENGUIN 1985)

KIERKEGARD, SÖREN *SICKNESS UNTO DEATH* (PENGUIN 1989)

KLINE, MORRIS *MATHEMATICS IN WESTERN CULTURE* (SBG 1965)

KOESTLER, ARTHUR *THE ACT OF CREATION* (PICADOR 1969)

KOESTLER, ARTHUR *THE GHOST IN THE MACHINE* (HUTCHINSON 1967)

KOESTLER, ARTHUR *JANUS* (HUTCHINSON 1978)

KOESTLER, ARTHUR *THE SLEEPWALKERS* (PENGUIN 1986)

KOESTLER, ARTHUR *THE WATERSHED* (HEINEMEN 1960)

KOYRÉ, A. *NEWTONIAN STUDIES* (HARVARD 1965)

KÜBLER-ROSS ELISABETH *DEATH AND DYING* (MACMILLAN 1969)

KUHN, THOMAS *THE STRUCTURE OF SCIENTIFIC REVOLUTIONS* (UNIVERSITY OF CHICAGO PRESS 1996)

LAING, RD *THE DIVIDED SELF* (PENGIUN 1979)

LAING, RD *KNOTS* (RANDOM HOUSE 1972)

LAING, RD *THE POLITICS OF EXPERIENCE AND THE BIRD OF PARADISE* (PENGUIN 1979)

LEE, H.D.P *PLATO THE REPUBLIC* (PENGUIN 1973)

LÉVI-STRAUSS, CLAUDE *STRUCTURAL ANTHROPOLOGY* (DOUBLEDAY 1967)

LEWIS, C.S. *THE ABOLITION OF MAN* (HARPERCOLLINS 2001)

LEWIS, C.S. *CHRISTIAN REFLECTIONS* (HARPERCOLLINS 1979)

LEWIS, C.S *MERE CHRISTIANITY* (HARPERCOLLINS 1984)

LEWIS, C.S. *SURPRISED BY JOY* (FONTANA 1968)

LEWIS, C.S. *THE GREAT DIVORCE* (HARPERCOLLINS 1979)

LOCKE, JOHN *AN ESSAY CONCERNING HUMAN UNDERSTANDING* (FONTANA 1964)

LOSEE, JOHN *AN HISTORICAL INTRODUCTION TO THE PHILOSOPHY OF SCIENCE* (OXFORD 1988)

LOVEJOY, A.O. *THE GREAT CHAIN OF BEING* (CAMBRIDGE MASS 1936)

LOVELOCK, JAMES *GAIA: A NEW LOOK AT LIFE ON EARTH* (OXFORD 1979)

MANDROU, ROBERT *FROM HUMANISM TO SCIENCE 1480-1700* (THE HARVESTER PRESS 1979)

MARCUSE, HERBERT *ONE DIMENSIONAL MAN* (ABACUS 1972)

MASCALL, E.L. *EXISTENCE AND ANALOGY* (LONGMANS, GREEN AND CO LTD 1949)

McKIBBEN, BILL *THE END OF NATURE* (VIKING 1990)

•152

MERCHANT, CAROLYN *THE DEATH OF NATURE, WOMEN, ECOLOGY, AND THE SCIENTIFIC REVOLUTION* (WILDWOOD HOUSE 1982)

NAGEL, E. *THE STRUCTURE OF SCIENCE* (HARCOURT, BRACE & WORLD 1961)

NEEDHAM, JOSEPH *SCIENCE AND CIVILIZATION IN CHINA VOL.2* (CUP 1962)

OLSON, RICHARD *SCIENCE DEIFIED AND SCIENCE DEFIED* VOLUME 1 (BERKELEY UCP 1982) AND VOLUME 2 (BERKELEY UCP 1983)

PASCAL, BLAISE *PENSÉES* (PENGUIN 1966)

PEPPER, D. *THE ROOTS OF MODERN ENVIRONMENTALISM* (CROOM HELM 1984)

PICK, DANIEL *WAR MACHINE* (YALE 1993)

PLATO *THE REPUBLIC* (PENGUIN 1973)

POLANYI, MICHAEL & PROSCH, HARRY *MEANING* (UNIVERSITY OF CHICAGO PRESS 1976)

POLKINGHORNE, JC *THE QUANTUM WORLD* (PENGIUN 1990)

POPPER, KARL *OBJECTIVE KNOWLEDGE* (CLARENDON PRESS 1972)

ROSZAK, THEODORE *THE MAKING OF A COUNTER CULTURE* (FABER AND FABER 1976)

RUPERT HALL, A. *FROM GALILEO TO NEWTON* (DOVER 1981)

RUSSELL, BERTRAND *THE HISTORY OF WESTERN PHILOSOPHY* (UNWIN 1979)

RUSSELL, BERTRAND *THE WISDOM OF THE WEST* (DOUBLEDAY 1964)

DE SANTILLANA, GIORGIO *THE CRIME OF GALILEO* (UNIVERSITY OF CHICAGO 1955)

SCRUTON, ROGER *SPINOZA* (OXFORD 1986)

SHELDRAKE, RUPERT *THE PRESENCE OF THE PAST* (COLLINS 1988)

SHELDRAKE, RUPERT *A NEW SCIENCE OF LIFE* (PALADIN 1983)

SCHRÖDINGER, ERWIN *WHAT IS LIFE? MIND AND MATTER* (CAMBRIDGE 1967)

SCHRÖDINGER, ERWIN *SCIENCE AND HUMANISM* (CAMBRIDGE 1951)

SOLLOWAY, FRANK J. *FREUD, BIOLOGIST OF MIND* (FONTANA 1980)

TARNAS, RICHARD *THE PASSION OF THE WESTERN MIND* (PIMLICO 1995)

TAWNEY, R.H. *RELIGION AND THE RISE OF CAPITALISM* (PENGIUN 1961)

TAYLOR, A.E. *PLATO* (UNWIN 1960)

TOYNBEE, ARNOLD *A STUDY IN HISTORY* (OXFORD UNIVERSITY PRESS 1957)

TURNBULL, GRACE THE ESSENCE OF PLOTINUS (OXFORD 1934)

WHITE, MICHAEL *ISAAC NEWTON THE LAST SORCERER* (FOURTH ESTATE 1998)

WHITE, R.J. *EUROPE IN THE SEVENTEENTH CENTURY* (MACMILLAN 1965)

WHITEHEAD, A.N. *SCIENCE AND THE MODERN WORLD* (CAMBRIDGE UNIVERSITY PRESS 2011)

WILLEY, BASIL *THE SEVENTEENTH CENTURY BACKGROUND* (ARK 1986)

WILLIAMS, RAYMOND *CULTURE AND SOCIETY 1780-1950* (PELICAN 1964)

WOLFE, D. (EDITOR) *THE COMPLETE PROSE WORKS OF JOHN MILTON* (YALE UNIVERSITY PRESS 1939)

ZAJONC, ARTHUR *CATCHING THE LIGHT* (DOUBLEDAY BANTAM 1993)

Appreciate the Eclipse?
Please leave a review on Amazon

This book is updated regularly
-thoughts and feedback always appreciated-

contactgough@gmail.com

About

John P Gough

Born: 1964 Sewell mine, Rancagua, Chile - father was a geologist

Travels – childhood - moving around Latin America
        adulthood - just about everywhere (apart from the East)
Occasional freelance journalist eg. New Internationalist

Favourite Philosopher: Plotinus
        Author: Tolstoy (or maybe Raymond Chandler)
        Contemporary thinker: Owen Barfield
Favourite saying: 'Hurry is not of the devil, hurry is the devil' (Jung)

- Hates:

        Processed thought and cleverness
        Naive science
        Morons and Political Correctness
        Tacky contemporary narratives eg. 911

+ Loves:

        all things organic (real food, normal stuff, before adulterated)
        Aria dreadnought guitar
        collecting typewriters
        reading in the bath
            jumping in frozen pools
            fuchsias and kettlebells

++ Loves: my Chilean wife (love at first sight) & children Thomas and Sophia.

My mother says I 'always wanted to know' – that makes me curious, prone to
mystical experience (I've had my moments) and philosophical wanderings.
Special thanks – Roy Porter, now sadly deceased, who encouraged me in all this.

Regrets: never having been really creative eg. written a novel (perhaps one day)
probably because of the below (Cartesian education) killed my imagination (that's
my excuse anyway):

BA Hons Communication Studies - 1987 Liverpool University
PGCE Leicester University 1991 Communications in Higher and Further Education
Phd. Philosophy of Science 1993 Leicester University
 (actually a fail but like to mention it)

Made in the USA
Monee, IL
07 July 2026

56552434R00085